T0185468

Neutron Scattering with a Triple-Axis Spectrometer

Neutron scattering is an extremely powerful tool in the study of elemental excitations in condensed matter. This book provides a practical guide to basic techniques using a triple-axis spectrometer.

Introductory chapters summarize useful scattering formulas and describe the components of a spectrometer, followed by a comprehensive discussion of the resolution function and focusing effects. Later chapters include simple examples of phonon and magnon measurements, and an analysis of spurious effects in both inelastic and elastic measurements, and how to avoid them. Finally, polarization analysis techniques and their applications are covered.

This guide will allow graduate students and post-docs, as well as experienced researchers new to neutron scattering, to make the most efficient use of their experimental time.

GEN SHIRANE received his doctorate in physics from the University of Tokyo in 1954. He has used neutron scattering extensively in studies of magnetism, ferroelectricity, lattice dynamics, and phase transitions.

STEPHEN SHAPIRO obtained his PhD in physics from Johns Hopkins University in 1969. He is particularly interested in applying neutron and X-ray scattering techniques to study structural and magnetic phase transitions in condensed matter systems.

JOHN TRANQUADA received his PhD in physics from the University of Washington (Seattle) in 1983. His work focuses on experimental studies of correlated electron systems, especially high-temperature superconductors and transition-metal oxides using neutron scattering and other techniques.

All three authors are currently senior scientists at the Brookhaven National Laboratory, New York.

Neutron Scattering
with a Triple-Axis Spectrometer
Basic Techniques

Gen Shirane, Stephen M. Shapiro, and John M. Tranquada

Brookhaven National Laboratory

CAMBRIDGE
UNIVERSITY PRESS

CAMBRIDGE UNIVERSITY PRESS
Cambridge, New York, Melbourne, Madrid, Cape Town, Singapore, São Paulo

Cambridge University Press
The Edinburgh Building, Cambridge CB2 2RU, UK

Published in the United States of America by Cambridge University Press, New York

www.cambridge.org
Information on this title: www.cambridge.org/9780521411264

First published 2002
This digitally printed first paperback version 2006

A catalogue record for this publication is available from the British Library

Library of Congress Cataloguing in Publication data
Shirane, G.
Neutron scattering with a triple-axis spectrometer / Gen Shirane, Stephen M. Shapiro,
and John M. Tranquada.
p.   cm.
ISBN 0 521 41126 2
1. Neutrons–Scattering.   I. Shapiro, S. M. (Stephen M.), 1941–
II. Tranquada, John M., 1955–   III. Title.
QC793.5.N4628 S47 2002
539.7'213–dc21   2001035050

ISBN-13 978-0-521-41126-4 hardback
ISBN-10 0-521-41126-2 hardback

ISBN-13 978-0-521-02589-8 paperback
ISBN-10 0-521-02589-3 paperback

Contents

Contents

Preface

The triple-axis spectrometer (TAS) remains arguably the most versatile instrument used for neutron scattering studies, and research using TASs has been of fundamental importance to many areas of condensed-matter science. The power provided by the flexibility of the TAS brings with it some challenges, and determining the optimal conditions for a particular experiment (often referred to as "finding the window") is a significant one. Various experimental pitfalls have also been identified over the years.

For 32 years (1965–96), triple-axis spectrometers at Brookhaven's High Flux Beam Reactor (HFBR) were the main tools of our Neutron Scattering Group. During that time, many basic techniques for both how and how not to perform experiments were recorded in group research memos. When new post-docs joined the group, they were expected to familiarize themselves with these memos; in addition, they were given a substantial list of references, including numerous research articles, to look up and read. The present book is intended to present the collection of basic techniques in a more-or-less coherent fashion. Besides graduate students and post-docs, we hope that the information presented here will also be useful to experienced researchers who are new to neutron scattering.

Of course, this book would have no purpose if it were not for Bertram Brockhouse and his invention of the TAS. We were extremely pleased when the winners of the Noble Prize in Physics were announced in 1994. The honoring of Brockhouse was long overdue, and it was very appropriate that he shared the prize with Clifford Shull, who did so many of the original experiments demonstrating the power and potential of neutron scattering.

Triple-axis spectrometry played a major role in the scientific success of the High Flux Beam Reactor (HFBR) at Brookhaven, and many researchers contributed to the science and to new developments in technique. In particular, we wish to recognize J. D. Axe, R. J. Birgeneau, and R. A. Cowley, who made many important contributions over the years. Others deserving special recognition are A. Kevey, who designed the spectrometers at the HFBR, and F. Langdon, W. Lenz, and R. Rothe, who were responsible for the operation of and improvements to the spectrometers.

Most of the issues that became subjects of the group research memos appeared in the course of the experimental program at the HFBR. Over the years, many prominent scientists performed experiments on the triple-axis instruments at the HFBR, and contributed in various ways to the accumulated experience. Besides those already named above, a partial list includes (in alphabetical order): G. Aeppli, J. Akimitsu, J. Als-Nielsen, K. Asai, P. Böni, D. J. Buttrey, M. F. Collins, R. Comès, D. E. Cox, R. Currat, B. Dorner, J. Eckert, Y. Endoh, Y. Fujii, C. Glinka, A. Goldman, H. Graf, M. Greven, H. Grimm, J. Harada, J. M. Hastings, K. Hirakawa, S. Hoshino, M. T. Hutchings, M. Iizumi, Y. Ishikawa, K. Kakurai, K. Katsumata, B. Keimer, J. K. Kjems, J. E. Lorenzo, J. W. Lynn, C. F. Majkrzak, J. L. Martinez, M. Matsuda, S. Mitsuda, D. E. Moncton, H. A. Mook, K. Motoya, A. H. Moudden, Y. Noda, L. Passell, W. Press, R. Pynn, D. Richter, S. K. Satija, M. Sato, S. K. Sinha, C. Stassis, O. Steinsvoll, E. C. Svensson, K. Tajima, Y. J. Uemura, S. A. Werner, Y. Yamada, and H. Yoshizawa.

Many of the examples used in the book are taken from work done at the HFBR. In a few cases, new measurements have been made for demonstration purposes. We would especially like to thank P. M. Gehring, K. Hirota, and K. Yamada for help with these measurements, and for other assistance in preparing the text. We would also like to thank M. Hase, I. Zaliznyak, and A. Zheludev for critical comments and suggestions on several chapters; A. Zheludev also supplied the figure used on the cover. Finally, thanks go to our editor, S. Capelin, for not giving up on us over the decade it took to complete this manuscript.

Gen Shirane
Stephen Shapiro
John Tranquada
Brookhaven

1

Introduction

In this introductory chapter we discuss some of the properties of the neutron and how it interacts with matter. We compare steady-state reactors to pulsed spallation neutron sources. After a review of the scattering geometry of an experiment we discuss the various instruments used for inelastic scattering.

1.1 Properties of thermal neutrons

The neutron is an ideal probe with which to study condensed matter. It was first discovered in 1932, and four years later it was demonstrated that neutrons could be Bragg diffracted by solids (see Bacon, 1986).† The early experimenters struggled with the low flux from Ra–Be sources but, nevertheless, established the fundamentals of neutron diffraction. The future was assured by the construction of the first "atomic pile" by Fermi and his co-workers in 1942. Subsequent reactors produced more and more neutrons and the latest generation of high-flux reactors built in the 1960s and early 1970s produce a copious number of neutrons, making inelastic scattering studies practical. It is now widely accepted that neutron scattering is one of the most important and versatile techniques for probing condensed matter. This fact was recognized by the awarding of the Nobel Prize in Physics in 1994 to Profs. C. Shull of Massachusetts Institute of Technology (MIT) and B. Brockhouse of McMaster University (Canada) for their seminal contributions to the fields of elastic and inelastic neutron scattering. Arguably the most important instrument used in neutron spectroscopy is the triple-axis spectrometer (invented by Brockhouse, 1961) since it allows for a controlled measurement of the scattering function $S(\mathbf{Q}, \omega)$ at essentially any point in momentum ($\hbar\mathbf{Q}$) and energy ($\hbar\omega$) space. (Here $\hbar$ is Planck's constant divided

---

† References are located at the end of each chapter.

1

Table 1.1. *Properties of the neutron.*

Quantity	Value
Rest mass, m_n	1.675×10^{-24} g
Spin	$\frac{1}{2}$
Magnetic moment, μ_n	1.913 nuclear magnetons, μ_N
Charge	0

by 2π, ω is 2π times the frequency ν, and the wave vector $\mathbf{Q}$ will be defined further in §1.3.)

Several of the properties of the neutron are listed in Table 1.1. The mass of the neutron is nearly that of the proton, and this relatively large mass has several important consequences. From the production viewpoint, the energetic neutrons produced by fission or spallation of heavy nuclei can be slowed down, or moderated, by collisions with atoms of similar mass, such as hydrogen or deuterium. The resultant energy distribution of the moderated neutrons is Maxwellian, with the average velocity determined by the temperature of the moderating medium. Another consequence of the large neutron mass is that "thermal" neutrons (i.e., those neutrons coming from a moderator near room temperature) have energies in an appropriate range (1–100 meV) for studying a wide variety of dynamical phenomena in solids and liquids.

The zero net charge of the neutron means that it interacts very weakly with matter and penetrates deeply into a sample. It also easily transmits through sample enclosures used to control the environment. This is a very important experimental convenience that is unique to the neutron probe. Even more important is the fact that with zero charge there is no Coulomb barrier to overcome, so that the neutrons are oblivious to the electronic charge cloud and interact directly with the nuclei of atoms. The theory of the interaction between a neutron and the nucleus of an atom is still incomplete, but it is known to be very short range ($\sim 10^{-13}$ cm $= 1$ fm, see Appendix 1). Since this is much less than the wavelength of thermal neutrons, the interaction can be considered nearly point-like. Therefore, neutron–nucleus scattering contains only s-wave components, which implies that the scattering is isotropic and can be characterized by a single parameter, b, called the scattering length. The typical value of the scattering length for the elements is on the order 1×10^{-12} cm, comparable to the nuclear radius. (The scattering cross section for a single nucleus is equal to $4\pi b^2$.) The scattering length can be complex; however, the imaginary part only becomes

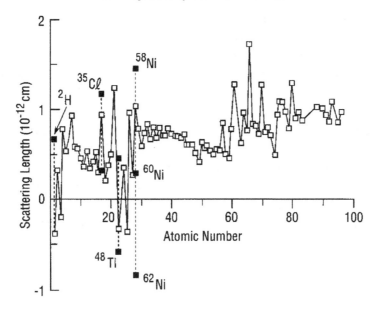

Fig. 1.1. Variation of coherent scattering length with atomic number (open squares). Variation of scattering length among isotopes of the same element is indicated for several cases by the filled squares. The coherent scattering length for an element corresponds to the average over all its isotopes, weighted by their abundance; the variations result in incoherent scattering, as discussed in Chap. 2 (from Price and Sköld, 1986).

significant near a nuclear absorption resonance. The real part is typically positive, but it can become negative at energies below a resonance; the sign of the scattering length is associated with the relative phase of the scattered neutron wave with respect to the incident wave. For most elements, the scattering length in the thermal energy regime is essentially independent of energy.

Figure 1.1 shows the variation of the scattering lengths for the different elements (Price and Sköld, 1986). (Appendix 1 tabulates these values.) One can see that most fall in the range between 0.2 and 1×10^{-12} cm. There is no systematic variation across the periodic table as there is for the case of X-ray or electron scattering. It is also seen that different isotopes of the same element can have different values of the scattering length, and can even be of different sign. The imaginary part of the scattering length represents absorption (principally radiative capture by the nuclei) and in most cases is small. Appendix 1 also lists the absorption cross sections for the elements. This quantity increases as the neutron energy approaches a nuclear resonance, so values must be quoted for a specific wavelength. Thus, in contrast with electrons and X-rays, the scattering does not depend upon the number of electrons and its strength varies somewhat randomly among the elements, and among isotopes of the same element. Also, because of the weakness of the interaction, the scattering amplitude from a sample is equal to the sum of the scattering amplitudes of the individual atoms. This result simplifies the interpretation of measurements.

Although the neutron carries no net charge, its internal structure of quarks and gluons gives the neutron a magnetic moment. The neutron's magnetic moment, listed in Table 1.1, interacts with the unpaired electron spins in magnetic atoms with a strength comparable to that of the nuclear interaction. The neutron is, therefore, a powerful probe of magnetic properties of solids and has contributed enormously to our understanding of magnetism.

The effective magnetic scattering length, p, for a magnetic atom with a moment of n_{μ_B} Bohr magnetons is given by:

$$p = 0.27 \times 10^{-12} \times n_{\mu_B} \text{ cm.} \tag{1.1}$$

The constant is one-half of the classical radius of the electron multiplied by the gyromagnetic ratio of the neutrons. For $n_{\mu_B} \gtrsim 1$, it is apparent that p is comparable to the b values shown in Fig. 1.1 or, in other words, the scattered intensity associated with magnetic effects is comparable to the scattering from the nuclei.

With a spin angular momentum of $\pm\frac{1}{2}\hbar$ per neutron, neutron beams can be prepared which contain a single angular momentum state, either spin up $(+\frac{1}{2}\hbar)$ or spin down $(-\frac{1}{2}\hbar)$. These spin-polarized neutrons have unique applications in determining magnetic structures, separating magnetic from nuclear scattering, and isolating incoherent scattering from the total scattering.

1.2 Neutron sources

The neutrons used in a scattering experiment can be obtained from a nuclear reactor (Table 1.2), where the neutrons arise from the spontaneous fission of ^{235}U, or from a spallation source (see Table 1.3) where the neutrons are produced by bombarding a heavy target (e.g., U, W, Ta, Pb, or Hg) with high-energy protons (Windsor, 1981). In the former case, the neutrons are produced continuously in time, while in the latter, they typically come as pulses. The continuous flux of the current generation of high-flux reactors [High Flux Isotope Reactor (HFIR) at Oak Ridge National Laboratory, and High Flux Reactor (HFR) at the Institute Laue–Langevin (ILL), Grenoble] is $\sim 1 \times 10^{15}$ neutrons/cm^2 s. The first such reactor to be optimized for neutron-beam research was the High Flux Beam Reactor (HFBR), see Fig. 1.2, at Brookhaven National Laboratory (BNL), which began operation in 1965. Although it is no longer operating, its design is typical of most high-flux reactors. In such a facility, the reactor, its auxiliary equipment, and the experimental facilities are contained in a sealed vessel of about 50 meters in diameter. This structure provides the final confinement against the escape of radioactive material into the environment. While the reactor is in operation the air pressure inside the building is kept slightly lower than atmospheric pressure outside to insure that any leakage is inward rather than outward. Access to the building is provided by a system of air locks.

Figure 1.2(b) shows a view of a segment of the experimental floor at the HFBR when it was operational. Neutrons from the reactor core were thermalized by a moderator of heavy water (D_2O); the thermalized neutrons reached the experimental floor through nine horizontal beam tubes. There were a total of 15 experimental facilities serving a wide variety of users. The newer neutron facilities, such as the HFR at ILL, Grenoble; Orphée at Laboratoire Léon Brillouin, Saclay; and the NIST Center for Neutron Research (NCNR) at the National Institute for Science and Technology near Washington, D.C., each have one or two "cold" sources. A cold source is a special moderator, typically utilizing liquid H_2 or CH_4 at cryogenic temperatures (~ 20 K), which shifts the peak of the Maxwellian distribution of neutrons to lower energy. The cold neutrons emerge in several beams; the separation between beams needed to allow for multiple instruments is achieved by transporting the neutrons over a long distance along slightly diverging paths into a "guide hall". For example, the HFR at ILL has, altogether, over 40 experimental facilities serving users in many disciplines.

At the pulsed sources (Table 1.3) neutrons are produced in bursts of roughly 10^{14} particles, with an initial pulse width on the order of $1\,\mu s$ at a frequency of 10–50 Hz. The time-averaged, moderated flux at existing pulsed sources is considerably less than at steady-state, reactor-based sources; however (Table 1.4), the 2 MW Spallation Neutron Source (SNS) being built in Oak Ridge will be an order of magnitude more powerful than the most intense existing pulsed (non-fission) spallation source, ISIS. [An even more powerful source, the European Spallation Source (ESS) is under consideration in Europe.] The SNS time-averaged flux will be approximately one-third that of the most powerful research reactor, the HFR at ILL; however, the peak flux from a spallation source is much higher. Efficient use of the neutron-beam time structure compensates for the lower time-averaged flux (Windsor, 1981).

The energy spectra from the two types of sources are slightly different, as

Table 1.2. *Operating research reactors with flux $\geq 10^{14}$ neutrons/(cm² s), from the data base of the International Atomic Energy Agency (http://www.iaea.org/worldatom/rrdb/).*

Name	Institution	Location	Country	Year of Startup	Power (MW)
NBSR	National Institute of Standards and Technology (NIST)	Gaithersburg, MD	USA	1969	20
HFIR	Oak Ridge National Laboratory (ORNL)	Oak Ridge, TN	USA	1966	85
MURR	University of Missouri	Columbia, MO	USA	1966	10
HFBR	Brookhaven National Laboratory (BNL)	Upton, NY	USA	1965[a]	60
NRU	Chalk River Laboratories	Chalk River, Ontario	Canada	1957	120
HFR	Institute Laue Langevin (ILL)	Grenoble	France	1972[b]	58
Orphée	Commissariat a l'Energie Atomique (CEA)	Saclay	France	1980	14
BER-II	Hahn-Meitner Institute	Berlin	Germany	1973[b]	10
FRJ-2	Forschungszentrum Jülich	Jülich	Germany	1962	23
BRR	Budapest Neutron Center	Budapest	Hungary	1959[b]	10
WWR-M	Petersburg Nuclear Physics Institute (PNPI)	Gatchina	Russia	1960	18
R-2	Neutron Research Lab (NFL)	Studsvik	Sweden	1960	50
DR3	Risø National Laboratory	Risø	Denmark	1960[c]	10
MARIE	Institute of Atomic Energy	Swierk	Poland	1974	30
CFANR	Nuclear Research Institute	Rez	Czech Republic	1957	10
HFR	Interfaculty Reactor Institute (IRI)	Delft	Netherlands	1961	45
JRR3M	Japanese Atomic Energy Research Institute (JAERI)	Tokai	Japan	1990	20
HANARO	Korea Atomic Energy Research Institute (KAERI)	Taejon	Korea	1996	30
HIFAR	Australian Nuclear Science & Technology Organization	Lucas Heights	Australia	1958	10
MRR	National Nuclear Energy Agency	Tangerang	Indonesia	1987	30
HWRR-II	Institute of Atomic Energy	Beijing	China	1958	15
DHRUVA	Bhabha Atomic Research Centre	Trombay	India	1985	100

[a] Permanently shut down in 1999.
[b] Upgraded.
[c] Permanently shut down in 2000.

Table 1.3. *Operating spallation sources.*

Name	Institution	Location	Country	Year of Startup	Power (kW)
IPNS	Argonne National Laboratory	Argonne, IL	USA	1981	7
LANSCE	Los Alamos National Laboratory	Los Alamos, NM	USA	1985	180
ISIS	Rutherford–Appleton Laboratory	Oxfordshire	England	1985	160
IBR-2	Joint Institute of Neutron Research (JINR)	Dubna	Russia	1984	2000[a]
SINQ	Paul Scherrer Institute	Villigen	Switzerland	1996	1000[b]
KENS	High-Energy Laboratory (KEK)	Tsukuba	Japan	1980	3

[a] Fission.
[b] Continuous.

(a)

(b)

Fig. 1.2. (a) Photograph of domed building of Brookhaven's High Flux Beam Reactor and (b) a view of the experimental floor of HFBR during full operations.

Table 1.4. *New neutron sources.*

Name	Institution	Location	Country	Type	Power (MW)	Status
SNS	Oak Ridge National Laboratory	Oak Ridge, TN	USA	Pulsed	2	Under construction
FRM-II	Technical University of Munich	Garching	Germany	Reactor	20	Awaiting startup approval
AUSTRON			Austria	Pulsed	0.5	Under consideration
CNF	Chalk River Laboratories	Chalk River, ON	Canada	Reactor	40	Under consideration
ESS	European Science Foundation	Undecided	Europe	Pulsed	5	Under consideration
CARR	China Institute of Atomic Energy	Undecided	China	Reactor	60	Under consideration
RRR	ANSTO	Lucas Heights	Australia	Reactor	20	Under construction
JOINT PROJECT	JAERI	Tokai	Japan	Pulsed	2	Under construction
PIK	PNPI	Gatchina	Russia	Reactor	100	Awaiting completion

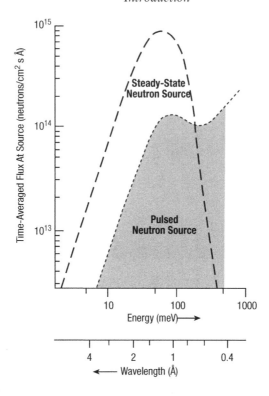

Fig. 1.3. Comparison of the source spectra of a steady-state reactor and a pulsed neutron source (from Windsor, 1981).

shown schematically in Fig. 1.3. For both types of sources the neutrons have to be moderated in order to reduce the effective neutron temperature to near room temperature. In a reactor, the flux is nearly isotropic and approximates a Maxwellian distribution of the neutron velocities with characteristic temperature of $\sim 350\,$K. The typical useful energy range of "thermal" neutrons is 5–100 meV; "cold" neutrons from a cryogenic moderator are roughly in the range of 0.1–10 meV. There is a relatively small epithermal ($> 100\,$meV) contribution, and this is beneficial for achieving a low background in the thermal regime. A spallation source has a much higher epithermal content in the energy spectrum, and the pulsed sources have a distinct advantage in exploiting this energy region. At a reactor, a useful flux in this higher-energy range can be provided with a "hot" source; the hot moderator at ILL consists of a block of graphite heated to 2400 K.

The pulsed beam from a spallation source is well suited to time-of-flight techniques. In one type of spectrometer, the incident neutrons are monochromatized using one or more choppers, and the energies of scattered neutrons are determined by the time it takes them to reach the detector. With the use of large arrays of position-sensitive detectors, it is possible to measure large regions of energy transfer, $\hbar\omega$, and momemtum transfer, $\hbar\mathbf{Q}$, simultaneously. At a reactor, neutron energy selection is more commonly achieved with Bragg-diffraction optics. The ability to focus neutrons onto the sample with a curved monochromator provides advantages for studies concentrating on a small region of ω–$\mathbf{Q}$ space.

1.3 Reciprocal space and scattering diagram

The laws of momentum and energy conservation governing all diffraction and scattering experiments are well known:

$$\mathbf{Q} = \mathbf{k}_f - \mathbf{k}_i \quad \text{(momentum conservation)} \tag{1.2}$$

$$|\mathbf{Q}| = k_i^2 + k_f^2 - 2k_i k_f \cos\theta_S \tag{1.3}$$

$$\hbar\omega = E_i - E_f \quad \text{(energy conservation).} \tag{1.4}$$

In these equations, the wave-vector magnitude $k = 2\pi/\lambda$, where λ is the neutron wavelength of the neutron beam, and the momentum transferred to the crystal is $\hbar\mathbf{Q}$. The subscript i refers to the beam incident on the sample and f the final or diffracted beam. The angle between the incident and final beams is $2\theta_S$ and the energy transferred to the sample is $\hbar\omega$. It is important to note that our sign convention for $\mathbf{Q}$ is the same as that used by Bacon (1975) but opposite to that of Lovesey (1984), who defines $\mathbf{Q}$ as $\mathbf{k}_i - \mathbf{k}_f$. The sign choice has implications for the resolution function discussed in Chap. 4.

Because of the finite mass of the neutron the dispersion relation for the neutron is:

$$E = \frac{\hbar^2 k^2}{2m_n},$$

$$E[\text{meV}] = 2.072k^2[\text{Å}^{-2}], \tag{1.5}$$

and the energy conservation law can be written as

$$\hbar\omega = \frac{\hbar^2}{2m_n}(k_i^2 - k_f^2). \tag{1.6}$$

Energy units are usually measured in meV, or equivalently as E/h in terahertz (THz $= 10^{12}$ Hz). Table 1.5 gives the energy relationships for neutrons in various units and their values at 10 meV. In this monograph we shall use mainly meV energy units. Wavelength, λ, and wave vector, k, will be expressed in Å and Å^{-1} units, respectively, where $1\,\text{Å} = 10^{-8}$ cm $= 0.1$ nm.

Table 1.5. *Wavelength, frequency, velocity and energy relationships for*
neutrons.

Quantity	Relationship	Value at $E = 10\,\mathrm{meV}$
Energy	$[\mathrm{meV}] = 2.072k^2[\text{Å}^{-1}]$	10 meV
Wavelength	$\lambda[\text{Å}] = 9.044/\sqrt{E[\mathrm{meV}]}$	2.86 Å
Wave vector	$k[\text{Å}^{-1}] = 2\pi/\lambda[\text{Å}]$	2.20 Å^{-1}
Frequency	$\nu[\mathrm{THz}] = 0.2418E[\mathrm{meV}]$	2.418 THz
Wavenumber	$\nu[\mathrm{cm}^{-1}] = \nu[\mathrm{Hz}]/(2.998 \times 10^{10}\,\mathrm{cm/s})$	80.65 cm^{-1}
Velocity	$v[\mathrm{km/s}] = 0.6302\,k[\text{Å}^{-1}]$	1.38 km/s
Temperature	$T[\mathrm{K}] = 11.605E[\mathrm{meV}]$	116.05 K

In any scattering experiment one always measures the properties of the incident (i) and final (f) neutron beams and infers the momentum and energy transferred to the sample via Eqs. (1.2) and (1.6). Since thermal neutrons have energies similar to those of many excitation processes of interest in solids (see Fig. 1.4), a rather modest and easily achievable energy resolution of $\Delta E/E \sim 10\%$ is frequently sufficient to obtain useful results. This is quite different from the case of inelastic scattering with X-rays or electrons, where, because of the larger probe energy ($\sim 10\,\mathrm{keV}$ for X-rays), a comparable absolute uncertainty ΔE may require $\Delta E/E \sim 10^{-7}$.

1.3.1 Elastic scattering

For the moment, let us consider only elastic scattering in a crystalline solid, i.e., $|\mathbf{k}_i| = |\mathbf{k}_f| = k$. To understand diffraction and scattering measurements, it is necessary to deal with the reciprocal lattice of the solid. The dots in Fig. 1.5 represent a reciprocal lattice for a two-dimensional crystalline solid, with each point corresponding to a reciprocal-lattice vector. If we plot a circle with radius k on this diagram such that it passes through two points on the circle, one of which is the origin of reciprocal space, the condition for Bragg scattering from the crystal is satisfied. The circle is called the Ewald circle in two dimensions, or the Ewald sphere in three dimensions. In the diagram, $\mathbf{k}_i$ is the direction of the incident beam relative to the crystal and $\mathbf{k}_f$ is the direction of the diffracted beam. For the case satisfying the Bragg condition:

$$\mathbf{Q} = \mathbf{G} = \mathbf{k}_f - \mathbf{k}_i \qquad (1.7)$$

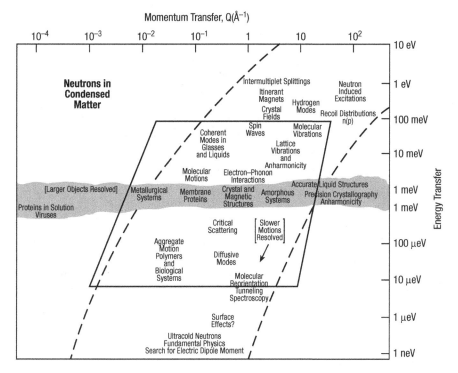

Fig. 1.4. Various applications of inelastic neutron scattering in terms of energy and momentum transfer. The region inside the trapezoidal box is the approximate range probed by three-axis instruments (from Lander and Emery, 1985). (The shaded region corresponds to an expanded cut in the energy scale.)

where $\mathbf{G}$ is a reciprocal-lattice vector. From Eq. (1.3) and Fig. 1.5

$$|\mathbf{Q}| = |\mathbf{G}| = 2|\mathbf{k}_i| \sin \theta_S, \qquad (1.8)$$

where $2\theta_S$ is the angle between the incident and the final beam for the Bragg condition. This is the well-known Bragg's law, which can also be written in the more familiar form

$$\lambda = 2d \sin \theta_S \qquad (1.9)$$

by noting that the magnitude of the reciprocal-lattice vector $|\mathbf{G}| = 2\pi/d$ where d is an interplanar spacing.

In a diffraction experiment, the magnitude of $\mathbf{Q}$ is controlled by adjusting the angle $2\theta_S$ between $\mathbf{k}_i$ and $\mathbf{k}_f$. The orientation of $\mathbf{Q}$ within the reciprocal lattice is set by rotating the sample. Thus, any point in reciprocal space can be measured by an appropriate choice of $\mathbf{k}_i$, $2\theta_S$, and the orientation ϕ of the sample relative to $\mathbf{k}_i$.

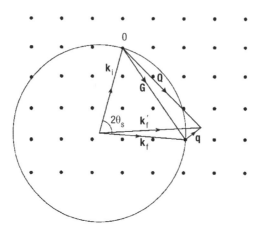

Fig. 1.5. Two-dimensional representation of reciprocal space showing the Ewald circle and the vector representation for elastic and inelastic scattering. Here **G** is a reciprocal-lattice vector and **q** the momentum transfer within the first Brillouin zone (see §1.3.2 and Fig. 1.6).

1.3.2 Inelastic scattering

For inelastic neutron scattering, the situation is more complicated. In this case $|\mathbf{k}_i| \neq |\mathbf{k}_f|$ since a difference is needed in order to transfer energy to the sample. In an experiment, one typically holds one wave vector (k_i or k_f) constant while varying the other. For a single-crystal sample, energies depend only on the relative momentum $\hbar\mathbf{q}$ defined within a Brillouin zone; hence, it is convenient to reference the momentum transfer to the nearest reciprocal lattice vector, i.e.,

$$Q = G + q, \tag{1.10}$$

as illustrated in Fig. 1.6. The measured spectrum can be interpreted straightforwardly if **Q** is held constant while the energy transfer is varied. Figure 1.6 shows two cases where k_i is kept constant and k_f varies. In the first situation [Fig. 1.6(a)], with $k_i > k_f$ and $\hbar\omega > 0$, energy is transferred from the incident neutron to the sample and an excitation is created; this is equivalent to Stokes scattering in optical spectroscopy. In the second example [Fig. 1.6(b)], $k_i < k_f$ and $\hbar\omega < 0$, so that the sample gives up a quantum of energy to the neutron beam. An excitation is annihilated and we have neutron energy gain (or anti-Stokes scattering).

In order to keep **Q**, and thus **q**, constant while varying $\mathbf{k}_f$, the scattering angle must change as well as the relative orientation of the crystal with respect to $\mathbf{k}_i$. The schematic in Fig. 1.6 shows the lattice staying fixed and

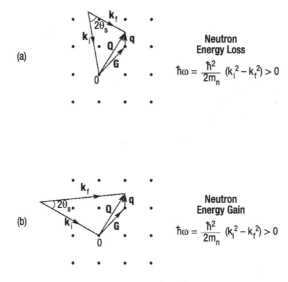

Fig. 1.6. Vector diagrams of inelastic scattering for (a) neutron energy loss ($k_f < k_i$), (b) neutron energy gain ($k_f > k_i$). 0 represents the origin of reciprocal space, **G** a reciprocal-lattice vector, and **q** the momentum transfer within a zone.

k_i moving. In practice, it is the other way around: k_i is kept fixed in space while the crystal is rotated.

1.4 Neutron scattering instruments

The main focus of this monograph is the triple-axis (or three-axis) spectrometer, which is an effective tool for both elastic and inelastic scattering studies. Here we compare it briefly with three other types of instrument commonly used for inelastic studies (see Fig. 1.7): time-of-flight (TOF), backscattering, and neutron spin-echo spectrometers.

1.4.1 Three-axis instrument

The three-axis instrument is the most versatile and useful instrument for use in inelastic scattering because it allows one to probe nearly any coordinates in energy and momentum space in a precisely controlled manner. The brilliant concept was developed over 40 years ago by Brockhouse (1961) at Chalk River in Canada. The three axes correspond to the axes of rotation of the monochromator, the sample, and the analyzer. The monochromator defines the direction and magnitude of the momentum of the incident beam and the analyzer performs a similar function for the scattered or final beam.

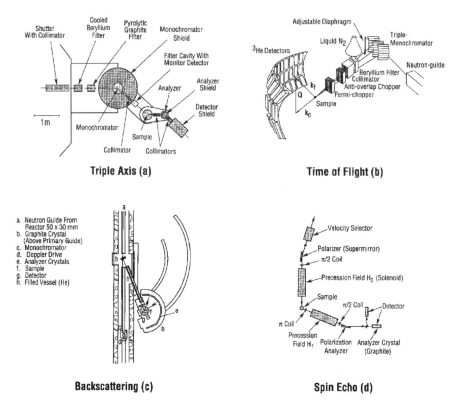

Fig. 1.7. Several types of spectrometers used at steady-state sources.

Figure 1.2(b) shows the three-axis instruments formerly situated at the H7 and H8 beam ports at BNL. Inside the large semi-cylindrical structure is the monochromator, and neutrons diffracted by it hit the sample. Those neutrons scattered by the sample are Bragg reflected by the analyzer, which lies within the white shielding. The instrument is large, mainly because of the necessary shielding which keeps the radiation levels low in the neighborhood of the instrument and keeps the background measured by the detector at the lowest possible level. The various components of the three-axis spectrometer, shown in Fig. 1.7(a), will be discussed in Chap. 3.

1.4.2 Time-of-flight (TOF) instrument

The time-of-flight technique was first employed to perform energy-dependent neutron measurements (Dunning et al., 1935). The technique has been greatly refined over the last 50 years with the advent of reactors and pulsed sources, but the principles remain the same. In the TOF technique, a burst of polychromatic neutrons is produced, and the times taken by the neutrons to travel from the source of the burst to the detector (counter) are measured. After interacting with the sample, the neutron will gain or lose energy, resulting in a velocity change. The arrival time at the counter will therefore vary.

If the neutron source is a reactor, a chopper or combination of choppers is used to define the initial pulse at time t_0. Frequently, instead of a chopper, a rotating crystal is used which simultaneously monochromatizes and pulses the neutron beam before it is scattered by the sample. Pulsed sources, by their nature, are ideally suited for TOF techniques. The pulse of polychromatic neutrons is provided by the accelerator, and the chopper, which is usually synchronized with the accelerator, defines the time t_0.

Figure 1.7(b) shows a schematic of the IN5 TOF spectrometer at ILL in Grenoble. This instrument is designed to give high energy resolution ($\sim 3\mu eV$) and is situated at a cold source. Two synchronized choppers are used to define the incident beam energy, while a third chopper is used to eliminate unwanted neutrons with velocities that are integral multiples of the desired neutron velocity. The fourth chopper spins at a lower speed in order to prevent frame overlap from different pulses. In this spectrometer the sample is placed in an argon-filled sample box allowing for a wide range of sample environments. The scattered neutrons then travel over a distance of about 4 meters to a number of detectors covering the angular range $-10° \leq 2\theta_S \leq 130°$. This instrument has a wide range of inelastic neutron scattering applications; notable among them are studies of tunneling states in hydrogen-containing materials.

1.4.3 Backscattering and neutron spin echo (NSE)

Figures 1.7(c) and 1.7(d) show two instruments used to achieve very high energy resolution. The backscattering instrument (Birr, Heidemann, and Alefield 1971; Alefeld, Springer, and Heidemann, 1992) uses very good crystals with narrow mosaics for monochromator and analyzer, each with a fixed Bragg angle of $\sim 90°$. Since one cannot scan in energy by varying the angle of the monochromator, the incident energy is instead tuned by a Doppler drive attached to the monochromator. Another method for scanning energy is to vary the temperature of the crystal, so that the lattice spacing of the monochromator is changed while keeping the scattering angle fixed. The new High Flux Backscattering Spectrometer at NCNR has achieved an energy resolution of $< 1\,\mu eV$ with a dynamic range greater than $\pm 40\,\mu eV$ (Gehring and Neumann, 1998).

The neutron spin echo (NSE) technique is a novel one developed (Mezei, 1972) to obtain high energy resolution without the concommitant intensity losses associated with backscattering geometries and the use of nearly perfect crystals. It relies on the use of polarized beams traversing a magnetic field region and the measurement of the number of precessions the neutron makes before and after hitting the sample. The schematic of an NSE spectrometer, the IN11 at ILL, is shown in Fig. 1.7(d). A cold source emits a neutron beam with a relatively broad band of energies which is then polarized by a supermirror. The spin of the neutron is flipped so that it is perpendicular to the magnetic field of the solenoid. The neutron will then make a number of Larmor precessions which depend upon the field strength and the length of the field region. The neutrons then strike the sample and gain or lose a small amount of energy. The spins are flipped 180° and enter a second precession field where they precess in the opposite direction. The spins are then flipped by 90° so they can be analyzed and detected. The difference between the number of precessions before and after the sample is proportional to the change in the velocity of the neutron after interacting with the sample. In this technique, since the precession of each neutron is measured, the resolution of the neutron energy transfer is independent of the beam monochromatization. The quantity measured is not the usual scattering function, $S(\mathbf{Q},\omega)$, but its time-dependent Fourier transform, $S(\mathbf{Q},t)$. If one compares the effective energy resolution by converting from the time domain, the NSE is able to achieve neV resolution, much better than that obtainable with more conventional instruments. This instrument is especially useful in probing the dynamics of macromolecular systems where large molecules move very slowly. It complements optical correlation spectroscopy by being able to probe a much larger range of momentum transfer.

References

Alefeld, B., Springer, T., and Heidemann, A. (1992). *Nucl. Sci. Eng.* **110**, 84.

Bacon, G. E. (1975). *Neutron Diffraction* (Clarendon Press, Oxford).

Bacon, G. E. (1986). *Fifty Years of Neutron Diffraction: The Advent of Neutron Scattering*, ed. G. E. Bacon (Adam Hilger, Bristol).

Birr, M., Heidemann, A., and Alefeld, B. (1971). *Nucl. Instrum. Methods* **95**, 435.

Brockhouse, B. N. (1961). In *Inelastic Neutron Scattering in Solids and Liquids* (International Atomic Energy Agency), p. 113.

Dunning, J. R., Pegram, G. B., Fink, G. A., Mitchell, D. P., and Segre, E. (1935) *Phys. Rev.* **49**, 704.

Gehring, P. M. and Neumann, D. A. (1998). *Physica B* **241–243**, 64.

Lander, G. H. and Emery, V. J. (1985). *Nucl. Instrum. Methods B* **12**, 525.

Lovesey, S. W. (1984). *Theory of Neutron Scattering from Condensed Matter* (Clarendon Press, Oxford).

Mezei, F. (1972). *Z. Phys.* **255**, 146.

Price, D. L. and Sköld, D. L. (1986). In *Methods of Experimental Physics, Vol. 23: Neutron Scattering*, ed. K. Sköld and D. L. Price (Academic Press, Orlando), Part A, p. 1.

Windsor, C. G. (1981). *Pulsed Neutron Diffraction* (Taylor Francis, London).

2

Scattering formulas

2.1 Introduction

Consider a monoenergetic beam of neutrons characterized by wave vector $\mathbf{k}_i$ and flux $\phi(k_i)$, incident on a sample. The rate at which they are scattered by a sample is given by the product $\phi(k_i)\sigma$, where σ is the scattering cross section. In a neutron scattering experiment, we are interested in the rate at which neutrons are scattered into a given solid angle $d\Omega_f$, in the direction of the wave vector $\mathbf{k}_f$, with a final energy between E_f and $E_f + dE_f$. This rate is given by the product of $\phi(k_i)$ and the double-differential cross section, $d^2\sigma/d\Omega_f dE_f$. It is convenient to express the differential cross section as a sum of coherent and incoherent parts:

$$\frac{d^2\sigma}{d\Omega_f dE_f} = \left.\frac{d^2\sigma}{d\Omega_f dE_f}\right|_{\text{coh}} + \left.\frac{d^2\sigma}{d\Omega_f dE_f}\right|_{\text{inc}}. \tag{2.1}$$

In general terms, the coherent part provides information about the cooperative effects among different atoms, such as elastic Bragg scattering or inelastic scattering by phonons or magnons, whereas the incoherent part is proportional to the time correlation of an atom with itself and provides information about individual particle motion, such as diffusion. The distinction between coherent and incoherent scattering will be made clearer below.

There are several excellent books (Bacon, 1975; Lovesey, 1984; Squires, 1978) which derive cross sections for various scattering processes in considerable detail and generality. In this chapter we will summarize some of the more useful formulas for interpreting experimental measurements on crystalline solids. For more general results and further details, the interested reader is encouraged to consult the references listed above.

2.2 Fermi's Golden Rule and the Born approximation

As far as probing condensed matter is concerned, a neutron acts as a very weak perturbation of the scattering system. When a neutron scatters, it can cause a transition of the sample from one quantum state to another, but it does not modify the nature of the states themselves. As a consequence, the differential scattering cross section can be obtained from Fermi's Golden Rule. Let V represent the interaction operator for the neutron with the sample; then, if the initial and final states of the sample are labeled by quantum numbers λ_i and λ_f, the differential cross section is

$$\frac{d^2\sigma}{d\Omega_f dE_f}\bigg|_{\lambda_i \to \lambda_f} = \frac{k_f}{k_i}\left(\frac{m_n}{2\pi\hbar^2}\right)^2 |\langle k_f \lambda_f | V | k_i \lambda_i \rangle|^2 \, \delta(\hbar\omega + E_i - E_f). \qquad (2.2)$$

Because the effective interaction is weak, one can evaluate the interaction matrix element using the Born approximation, treating both the incident and outgoing neutrons as plane waves:

$$\langle k_f \lambda_f | V | k_i \lambda_i \rangle = V(\mathbf{Q})\langle \lambda_f | \sum_l e^{i\mathbf{Q}\cdot\mathbf{r}_l} | \lambda_i \rangle, \qquad (2.3)$$

where $\mathbf{r}_l$ are the coordinates of the scattering centers (assumed here to be identical), and

$$V(\mathbf{Q}) = \int d\mathbf{r}\, V(\mathbf{r}) e^{i\mathbf{Q}\cdot\mathbf{r}}. \qquad (2.4)$$

For nuclear scattering, the nuclear potential is essentially a delta function in $\mathbf{r}$, so

$$V(\mathbf{Q}) = \frac{2\pi\hbar^2}{m_n} b, \qquad (2.5)$$

where b is the nuclear scattering length.

In a scattering experiment, one generally averages over initial states and sums over final states. If $P(\lambda_i)$ is the statistical weight factor for initial state $|\lambda_i\rangle$, then

$$\frac{d^2\sigma}{d\Omega_f dE_f} = \frac{k_f}{k_i} \sum_{\lambda_i,\lambda_f} P(\lambda_i) \left| \left\langle \lambda_f \left| b \sum_l e^{i\mathbf{Q}\cdot\mathbf{r}_l} \right| \lambda_i \right\rangle \right|^2 \delta(\hbar\omega + E_i - E_f). \qquad (2.6)$$

(We will generalize this later to include magnetic scattering.) Using a couple of standard tricks, Van Hove (1954) showed that this can also be expressed as

$$\frac{d^2\sigma}{d\Omega_f dE_f} = N\frac{k_f}{k_i} b^2 S(\mathbf{Q}, \omega), \qquad (2.7)$$

where

$$S(\mathbf{Q}, \omega) = \frac{1}{2\pi\hbar N} \sum_{ll'} \int_{-\infty}^{\infty} dt \left\langle e^{-i\mathbf{Q}\cdot\mathbf{r}_{l'}(0)} e^{i\mathbf{Q}\cdot\mathbf{r}_l(t)} \right\rangle e^{-i\omega t}, \tag{2.8}$$

N is the number of nuclei, t is time, and we have used the angle brackets, $\langle\ldots\rangle$, to denote the average over initial states. The scattering function depends only on the momentum and energy transferred from a neutron to the sample, and not on the absolute values of $\mathbf{k}_i$ and $\mathbf{k}_f$. It contains information on both the positions and motions of the atoms comprising the sample. The goal of most neutron scattering experiments is to measure $S(\mathbf{Q}, \omega)$ and thereby determine the microscopic properties of the system under investigation.

2.3 Coherent vs. incoherent scattering

To explain the difference between the coherent and incoherent cross sections, we begin by noting that a monatomic sample may contain nuclei with varying scattering lengths. Isotopes of the same element have different scattering lengths; furthermore, for an isotope with a nuclear spin, the scattering length varies depending on whether the neutron and nuclear spins are parallel or antiparallel. Suppose that the rth distinct isotope or nuclear spin state has scattering length b_r and occurs with frequency c_r. If there are no correlations between nuclear position and scattering length, then scattering which depends on the relative positions of the atoms will depend only on the average (or coherent) scattering length, given by

$$\bar{b} = \sum_r c_r b_r. \tag{2.9}$$

The average coherent cross section per atom is then

$$\sigma_{\text{coh}} = 4\pi \left(\bar{b}\right)^2. \tag{2.10}$$

The random fluctuations in scattering length from site to site will not contribute to collective scattering, but only to incoherent scattering. The total scattering cross section is given by

$$\sigma_{\text{scat}} = 4\pi \sum_r c_r b_r^2 = 4\pi \overline{b^2}, \tag{2.11}$$

so, using $\sigma_{\text{inc}} = \sigma_{\text{scat}} - \sigma_{\text{coh}}$, the incoherent cross section per atom is given by

$$\sigma_{\text{inc}} = 4\pi \left(\overline{b^2} - \bar{b}^2\right) = 4\pi \overline{(b - \bar{b})^2}. \tag{2.12}$$

An effective incoherent scattering length can be written as

$$b_{\text{inc}} = \sqrt{\overline{b^2} - \bar{b}^2}.\tag{2.13}$$

A table of the coherent and incoherent scattering lengths and cross sections for the elements is given in Appendix 1. Most elements have a significant coherent cross section, but there are a few prominent examples, such as hydrogen and vanadium, for which the incoherent scattering is large and dominant. The incoherent cross section is zero only for single isotopes with zero nuclear spin. It is sometimes desirable and possible to choose a material with a single isotope of zero spin such as ^{4}He, ^{36}Ar, ^{58}Ni, or ^{154}Sm in order to eleminate or minimize incoherent scattering.

2.4 Coherent nuclear scattering

Considering just the coherent scattering, the differential cross section is

$$\frac{d^2\sigma}{d\Omega_f dE_f}\bigg|_{\text{coh}} = N\frac{k_f}{k_i}\frac{\sigma_{\text{coh}}}{4\pi}S(\mathbf{Q},\omega).\tag{2.14}$$

This formula applies when the sample consists of a single element. More generally, one must include the site-dependent scattering lengths in the scattering function $S(\mathbf{Q},\omega)$.

An elegant way in which to write the scattering function was given by Van Hove (1954) using the definition of the atomic density operator:

$$\rho_{\mathbf{Q}}(t) = \sum_l e^{i\mathbf{Q}\cdot\mathbf{r}_l(t)}.\tag{2.15}$$

With this, one can write

$$S(\mathbf{Q},\omega) = \frac{1}{2\pi\hbar N}\int_{-\infty}^{\infty} dt\, e^{-i\omega t}\langle \rho_{\mathbf{Q}}(0)\rho_{-\mathbf{Q}}(t)\rangle.\tag{2.16}$$

Thus, the scattering function is the Fourier transform of the time-dependent pair-correlation function.

2.4.1 Elastic scattering (Bragg peaks)

For coherent, elastic nuclear scattering, we consider the time average of the density operator, so that

$$S(\mathbf{Q},\omega) = \delta(\hbar\omega)\frac{1}{N}\left\langle \sum_{ll'} e^{i\mathbf{Q}\cdot(\mathbf{r}_l-\mathbf{r}_{l'})}\right\rangle.\tag{2.17}$$

In the case of a Bravais lattice, this becomes

$$S(\mathbf{Q}, \omega) = \delta(\hbar\omega) \frac{(2\pi)^3}{v_0} \sum_{\mathbf{G}} \delta(\mathbf{Q} - \mathbf{G}), \tag{2.18}$$

where v_0 is the unit-cell volume and the vectors $\mathbf{G}$ are reciprocal-lattice vectors. The coherent, elastic cross section is then

$$\left. \frac{d\sigma}{d\Omega} \right|_{\mathrm{coh}}^{\mathrm{el}} = N \frac{(2\pi)^3}{v_0} (\bar{b})^2 \sum_{\mathbf{G}} \delta(\mathbf{Q} - \mathbf{G}). \tag{2.19}$$

This formula describes the scattering from a perfectly rigid lattice. In practice, the fluctuations of the atoms about their equilibrium positions causes some reduction of the Bragg intensities. Let $\mathbf{u}$ be the instantaneous displacement of an atom from its equilibrium position $\mathbf{r}$. Averaging of the phase factor $\exp(-i\mathbf{Q} \cdot \mathbf{r})$ in Eq. (2.17) results in an extra factor in Eq. (2.19) equal to $\exp(-2W)$ (known as the Debye–Waller factor), where, for small displacements,

$$W = \tfrac{1}{2} \left\langle (\mathbf{Q} \cdot \mathbf{u})^2 \right\rangle. \tag{2.20}$$

In the case of a cubic crystal, where the atomic displacements are the same in all directions,

$$\langle u_x^2 \rangle = \langle u_y^2 \rangle = \langle u_z^2 \rangle = \tfrac{1}{3} \langle u^2 \rangle, \tag{2.21}$$

so that

$$W = \tfrac{1}{6} Q^2 \langle u^2 \rangle. \tag{2.22}$$

Crystallographers tend to use a slightly different notation, in which

$$W = B \sin^2 \theta / \lambda, \tag{2.23}$$

with B typically assumed to be isotropic. It follows that

$$B = \frac{8\pi^2}{3} \langle u^2 \rangle. \tag{2.24}$$

Thus far, we have limited ourselves to the case of a Bravais lattice. Suppose, instead, that we are interested in a lattice with more than one atom per unit cell. If the jth atom within the unit cell sits at position $\mathbf{d}_j$, then the coherent elastic differential cross section generalizes to

$$\left. \frac{d\sigma}{d\Omega} \right|_{\mathrm{coh}}^{\mathrm{el}} = N \frac{(2\pi)^3}{v_0} \sum_{\mathbf{G}} \delta(\mathbf{Q} - \mathbf{G}) |F_{\mathrm{N}}(\mathbf{G})|^2, \tag{2.25}$$

where

$$F_N(\mathbf{G}) = \sum_j \bar{b}_j e^{i\mathbf{G}\cdot\mathbf{d}_j} e^{-W_j}. \tag{2.26}$$

The static nuclear structure factor $F_N(\mathbf{G}) = F_N(hkl)$ contains information on the atomic positions $\mathbf{d}_j$ within a unit cell and the mean-square displacements $\langle u_{j\alpha}^2 \rangle$. If one measures structure factors for a large number of reflections, a model for the atomic parameters can be fitted to the results; this is the standard approach of crystallography. The square of the nuclear structure factor can be obtained from a scan through a Bragg peak. The precise formula for the integrated intensity of a peak scan depends, in general, on the resolution function of the instrument; however, there are a couple of simple cases that are easily described.

To begin, consider a measurement on a single crystal using a monochromatic incident beam of wavelength λ. For a reflection (hkl) in the scattering plane, with scattering angle θ relative to the diffracting planes, the diffracted beam is deflected by 2θ with respect to the incident beam. If a detector is placed at 2θ, and then measurements are made as the crystal is rotated through the reflection, the area under the observed peak (integrated intensity) is given by

$$\mathscr{I} = A \frac{\lambda^3 |F_N(hkl)|^2}{v_0^2 \sin 2\theta}, \tag{2.27}$$

where A is a constant that depends on the incident flux, the sample volume, and the counting time. If these factors are all held fixed in a series of measurements, then A is simply an overall scale factor.

Note that we have described a measurement with no analyzer crystal or Soller collimation (see Chap. 3) after the sample; these are typically not used in a two-axis diffractometer. Using either or both of these changes the θ dependence of the formula, as discussed in Chap. 7; however, there is one mode of measurement with a three-axis spectrometer for which Eq. (2.27) still applies. That mode involves scanning the analyzer arm (2θ) in steps twice the size of those applied to the sample (θ), a so-called θ-2θ scan. The advantage of the 3-axis approach is that one can discriminate against the scattering from acoustic phonons that tends to be peaked at the positions of Bragg reflections. (The phonon contribution is often called thermal diffuse scattering, because the strength of the acoustic phonon signal has a significant temperature dependence.)

For a sufficiently large single crystal, the diffracted intensity for strong reflections will eventually saturate, so that the integrated intensity is no longer proportional to $|F_N|^2$. The precise relationship between the intensity

and $|F_N|^2$ will depend on the nature and shape of the crystal. A formula that
is sometimes useful for modeling the measured intensities is

$$\mathscr{I} \sin 2\theta = \frac{\alpha_1 |F_N^{calc}|^2}{1 + \alpha_2 |F_N^{calc}|^2}, \tag{2.28}$$

where F_N^{calc} is the calculated structure factor, and α_1 and α_2 are adjustable
parameters.

When the sample is polycrystalline (with randomly oriented grains), the
orientation of the sample is irrelevant, and a Bragg peak is measured by
rotating the detector. In this case, the appropriate formula for the integrated
intensity is

$$\mathscr{I} = A \frac{m_{hkl} \lambda^3 |F_N(hkl)|^2}{v_0^2 \sin\theta \sin 2\theta}, \tag{2.29}$$

where m_{hkl} is the multiplicity of the reflection (hkl). The multiplicity is the
number of equivalent permutations of the indices hkl; for example, the
multiplicity of a (100) reflection from a cubic crystal is 6.

2.4.2 Inelastic scattering

If we subtract out the elastic contributions such as Bragg scattering, then
$S(\mathbf{Q}, \omega)$ corresponds to the fluctuations in a sample, as a function of momentum and frequency. An important property of the scattering function is
the principle of detailed balance:

$$S(-\mathbf{Q}, -\omega) = e^{-\hbar\omega/k_B T} S(\mathbf{Q}, \omega), \tag{2.30}$$

where k_B is Boltzmann's constant, T is temperature, and ω is assumed to be
positive. This property expresses the fact that the probability of a transition
in the sample depends on the statistical weight factor for the initial state,
which will be lower for excitation annihilation than for excitation creation.

The scattering function is related to the dissipative part of a linear response
function via the fluctuation–dissipation theorem,

$$S(\mathbf{Q}, \omega) = \frac{\chi''(\mathbf{Q}, \omega)}{1 - e^{-\hbar\omega/k_B T}}, \tag{2.31}$$

where $\chi''(\mathbf{Q}, \omega)$ is the imaginary part of the dynamical susceptibility. Note
that this formula applies to both positive and negative values of ω; positive ω
corresponds to a transfer of energy from the neutron to the sample, whereas,
for negative ω, energy is transferred from the sample to the scattered neutron.
Note that in cases where $S(-\mathbf{Q}, \omega) = S(\mathbf{Q}, \omega)$, the detailed-balance principle

requires that $\chi''(\mathbf{Q}, \omega)$ be an odd function of ω; the thermal factor also changes sign at $\omega = 0$, so that $S(\mathbf{Q}, \omega)$ is always positive.

To gain a proper understanding of correlation and linear-response functions, the reader may turn to the classic paper by Marshall and Lowde (1968) or to the excellent presentation in Chap. 7 of Chaikin and Lubensky (1995). For our present purposes of summarizing useful formulas, it is sufficient to note that Eq. (2.31) is especially useful when $\chi''(\mathbf{Q}, \omega)$ is not explicitly dependent on temperature. Such is the case for phonons, which we consider next.

2.4.2.1 Phonons

Consider a lattice with n atoms per unit cell. Such a lattice will have a total of $3n$ distinct phonon branches with frequencies ω_{qs}, where s labels the different modes and the wave vector $\mathbf{q}$ is measured from the nearest reciprocal-lattice vector $\mathbf{G}$. In a scattering measurement one has

$$\mathbf{Q} = \mathbf{G} + \mathbf{q}. \tag{2.32}$$

For a particular mode s, the polarization vector for the jth atom in the unit cell is $\boldsymbol{\xi}_{js}$. For the case where a neutron scattering from the lattice creates or destroys a single phonon, one has

$$\chi''(\mathbf{Q}, \omega) = \frac{1}{2} \frac{(2\pi)^3}{v_0} \sum_{\mathbf{G}, \mathbf{q}} \delta(\mathbf{Q} - \mathbf{q} - \mathbf{G}) \sum_s \frac{1}{\omega_{qs}} |\mathscr{F}(\mathbf{Q})|^2$$
$$\times \left[\delta\left(\omega - \omega_{qs}\right) - \delta\left(\omega + \omega_{qs}\right) \right], \tag{2.33}$$

where the dynamic structure factor, $\mathscr{F}(\mathbf{Q})$, is given by

$$\mathscr{F}(\mathbf{Q}) = \sum_j \frac{\bar{b}_j}{\sqrt{m_j}} (\mathbf{Q} \cdot \boldsymbol{\xi}_{js}) e^{i\mathbf{Q} \cdot \mathbf{d}_j} e^{-W_j}. \tag{2.34}$$

In the latter formula, m_j is the mass of the jth atom. Note that the minus sign between the delta functions in Eq. (2.33) is a consequence of the requirement that χ'' be odd in ω.

As we will discuss in Chaps. 4 and 5, the typical way to measure a phonon is to work at a fixed wave vector $\mathbf{G} + \mathbf{q}$ and vary the frequency through a branch at $+\omega_{qs}$ or $-\omega_{qs}$. To evaluate the integrated intensity for such a scan, we need to make use of Eqs. (2.33), (2.34), (2.31), and

$$\frac{d^2\sigma}{d\Omega_f dE_f} = N \frac{k_f}{k_i} S(\mathbf{Q}, \omega). \tag{2.35}$$

In Chap. 4 we will show that the factor k_f/k_i cancels out in a standard triple-axis measurement; as a result, the integrated intensity for a constant-$\mathbf{Q}$ scan is given by

$$\mathscr{I} = A \frac{1}{\omega_{\mathbf{q}s}} |\mathscr{F}(\mathbf{Q})|^2 \times \begin{cases} n_{\mathbf{q}s} + 1 & \text{for neutron energy loss,} \\ n_{\mathbf{q}s} & \text{for neutron energy gain,} \end{cases} \tag{2.36}$$

where

$$n_{\mathbf{q}s} = \frac{1}{e^{\hbar\omega_{\mathbf{q}s}/k_B T} - 1} \tag{2.37}$$

is the Bose factor. Some useful limiting cases for the Bose factor are

$$n_{\mathbf{q}s} + 1 = \begin{cases} 1 & \text{for } \hbar\omega_{\mathbf{q}s} \gg k_B T, \\ k_B T/\hbar\omega_{\mathbf{q}s} & \text{for } \hbar\omega_{\mathbf{q}s} \ll k_B T. \end{cases} \tag{2.38}$$

Thus, for a neutron energy-loss measurement at high temperature, Eq. (2.36) becomes

$$\mathscr{I} = A \frac{k_B T}{\hbar\omega_{\mathbf{q}s}^2} |\mathscr{F}(\mathbf{Q})|^2. \tag{2.39}$$

The intensity of phonon scattering depends on $\mathscr{F}(\mathbf{Q})$, which, in turn, depends on the phonon polarization vectors. There is no simple, generic way to predict the $\mathbf{Q}$ dependence of the polarization vectors, although they can, in principle, be calculated from suitable force constant models. The situation simplifies for acoustic modes in the limit $\mathbf{q} \to 0$ (Ishikawa *et al.*, 1985). To see this, we first note that the displacement $\mathbf{u}_{js}$ of the jth ion in the sth mode is given by

$$\mathbf{u}_{js} = \frac{\boldsymbol{\xi}_{js}}{\sqrt{m_j}}, \tag{2.40}$$

and that, as a consequence of orthonormality, the polarization vectors satisfy the relationship

$$\sum_j |\boldsymbol{\xi}_{js}|^2 = 1. \tag{2.41}$$

For an acoustic mode ($s = a$) in the limit of small $|\mathbf{q}|$, all atomic displacements $\mathbf{u}_{ja}$ become equal, and using Eqs. (2.40) and (2.41) one can show that

$$\lim_{\mathbf{q}\to 0} \frac{\boldsymbol{\xi}_{ja}}{\sqrt{m_j}} = \frac{\hat{\mathbf{e}}}{M}, \tag{2.42}$$

where $\hat{\mathbf{e}}$ is a unit vector parallel to $\mathbf{u}_{ja}$, and

$$M = \sum_j m_j. \tag{2.43}$$

It follows that

$$\lim_{q \to 0} \mathscr{F}^2_{\text{acoustic}}(\mathbf{Q}) = \frac{G^2}{M} F_N(\mathbf{G}), \tag{2.44}$$

where $F_N(\mathbf{G})$ is the nuclear structure factor for the Bragg peak at $\mathbf{Q} = \mathbf{G}$. This result can be useful for providing a relative calibration of the cross section for inelastic scattering from other sources, such as magnons.

2.4.2.2 Damped harmonic oscillator

In real systems, phonon–phonon and electron–phonon interactions tend to give single phonons a finite lifetime. A natural way in which to take account of this dissipation is with the damped harmonic-oscillator model; for a complete discussion of the model, see Chaiken and Lubensky (1995). In brief, the effect of damping is accommodated by replacing the delta functions with lorentzians and simultaneously renormalizing the phonon frequencies. Equation (2.33) still applies if we make the substitution

$$\frac{1}{\omega_{qs}} \delta(\omega \pm \omega_{qs}) \to \frac{1}{\pi \omega'_{qs}} \frac{\Gamma_{qs}}{[\omega \pm \omega'_{qs}]^2 + \Gamma^2_{qs}}, \tag{2.45}$$

where Γ_{qs} is the peak half-width at half-maximum (HWHM), and

$$\omega'^2_{qs} = \omega^2_{qs} - \Gamma^2_{qs}. \tag{2.46}$$

Note that ω'_{qs} has a real solution only as long as $\omega_{qs} > \Gamma_{qs}$. Provided that Γ_{qs}/ω_{qs} is reasonably small, Eq. (2.36) for the integrated intensity will still be approximately correct when ω'_{qs} is substituted for ω_{qs}.

In the overdamped regime, $S(\mathbf{Q}, \omega)$ has only a single peak at the origin. For the case of strong overdamping, $\omega_{qs} \ll \Gamma_{qs}$, the following approximate formula applies:

$$S(\mathbf{Q}, \omega) = \frac{(2\pi)^3}{v_0} |\mathscr{F}(\mathbf{Q})|^2 \frac{k_B T}{\hbar \omega^2_{qs}} \frac{1}{\pi} \frac{\gamma_{qs}}{\omega^2 + \gamma^2_{qs}}, \tag{2.47}$$

where

$$\gamma_{qs} = \frac{\omega^2_{qs}}{2\Gamma_{qs}}, \tag{2.48}$$

and we have used the high-temperature expansion for the thermal factor. The latter approximation is generally appropriate, since the formula applies to a limited frequency range about $\omega = 0$.

Equation (2.47) is useful for describing the critical scattering that occurs at temperatures just above the critical temperature T_0 for a phase transition driven by a soft phonon mode. A classic example is the $\beta \to \alpha$ transition in quartz (SiO_2), which occurs at $T_0 = 846\,K$ (Axe and Shirane, 1970). In the high temperature β-phase the soft mode is overdamped and located at the Brillouin zone center. Figure 2.1(a) shows energy scans at such a position, $\mathbf{Q} = (1, 0, 3)$. The scattering from the soft mode appears as a broad lorentzian, easily separated from the sharp Bragg peak, with an intensity that grows rapidly as T_0 is approached. From Eq. (2.47), one expects that the intensity $\mathscr{I}$ obtained by integrating over ω should be proportional to $k_B T / \omega_0^2$, where ω_0 is the frequency of the soft mode. If follows that $T / \mathscr{I}$ should be proportional to ω_0^2; this quantity is plotted versus temperature in Fig. 2.1(b). The straight line through the data points is a fit to the equation

$$(\hbar\omega_0)^2 = a(T - T_c); \tag{2.49}$$

the fit yields $T_c = 836 \pm 1.8\,K$, 10 K less than the actual first-order transition temperature, T_0.

2.4.2.3 Eigenvector determination

From the $\mathbf{Q} \cdot \boldsymbol{\xi}$ factor in the dynamical structure factor $\mathscr{F}(\mathbf{Q})$ [Eq. (2.34)] it is possible to determine the eigenvectors of the normal modes by making an appropriate choice of scans about different reciprocal-lattice vectors, $\mathbf{G}$. For long-wavelength acoustic modes, where all atoms are moving in phase, the situation is particularly simple. For a longitudinal mode the atomic displacements are parallel to the direction of propagation, $\mathbf{q}$, while for a transverse mode $\boldsymbol{\xi} \perp \mathbf{q}$. A few examples are illustrated in Fig. 2.2. A scan at point A, with $\mathbf{q}$ parallel to [100] and $\mathbf{G} = (0, 1, 0)$, would detect a transverse acoustic mode. A scan at B, where $\mathbf{q}$, $\mathbf{G}$ and $\mathbf{Q}$ are all parallel to [100] would measure a longitudinal acoustic mode. Scans at C and D would measure longitudinal and transverse branches, respectively, along the $\langle 110 \rangle$ directions.

This approach can be extended to optic modes as well. In complicated crystals, one can reduce the number of independent parameters for polarization vectors along symmetry directions by applying group theory. The eigenvectors at the zone center can be written as a simple linear combination of symmetry coordinates:

$$\boldsymbol{\xi}_{js} = \sum_{\lambda} C_{\lambda j} \mathbf{s}_{j\lambda}, \tag{2.50}$$

where $C_{\lambda j}$ is the contribution of the λth symmetry coordinate, $\mathbf{e}_{j\lambda}$, to the eigenvector $\boldsymbol{\xi}_{js}$ of atom j moving according to phonon branch s. For optic modes, an additional condition that the center of mass does not move must also be satisfied:

$$\sum_{j} m_j \boldsymbol{\xi}_{js} = 0. \tag{2.51}$$

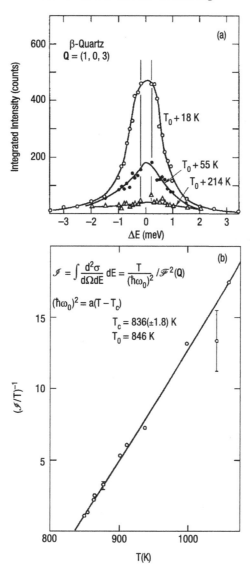

Fig. 2.1. Measurements of the overdamped soft mode at the $\beta \to \alpha$ transition in quartz by Axe and Shirane (1970). (a) Constant-$\mathbf{Q}$ scans at $\mathbf{Q} = (1, 0, 3)$ for several temperatures above the transition, $T_0 = 846$ K. (b) Inverse of the energy-integrated intensity $\mathscr{I}_E$ divided by temperature, plotted versus temperature. The straight line is a fit, as discussed in the text.

Brockhouse and Iyengar (1958) first demonstrated the use of eigenvector determination for germanium, which has the diamond structure. It has subsequently been applied to many other systems, including several with the perovskite structure (Harada, Axe, and Shirane 1970; Shirane, 1974). It has

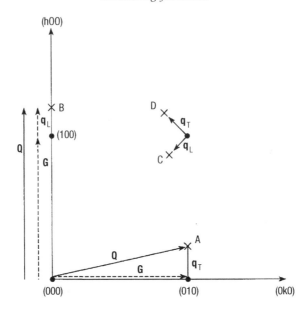

Fig. 2.2. Within an (*hk*0) zone of reciprocal space, purely transverse acoustic modes, subscript T, can be measured at points A and D, while purely longitudinal acoustic modes, subscript L are obtained at B and C.

also been applied to the study of the phase transition in quartz. Axe and Shirane (1970) determined the eigenvectors of the soft mode in the β-phase, which is both infrared- and Raman-inactive and so could only be studied by neutron scattering. The neutron experiment also unambiguously determined the eigenvector of the soft, zone-center mode in the low-temperature α-phase. This mode is a totally symmetric A_1 mode with an energy of 25.8 meV (208 cm^{-1}) at room temperature. Group theory states that there are four symmetry coordinates and three of these can be determined relative to the fourth. The intensity of the soft mode was measured about 18 reciprocal-lattice points, and the comparison between the calculated and observed intensities is shown in Fig. 2.3.

2.4.2.4 *Brillouin zones and intensity zones*

It is important to understand the difference between the Brillouin zone (BZ) and the intensity zone (IZ). The Brillouin zone is defined as a Wigner–Seitz cell in reciprocal space. It is the smallest unit in reciprocal space over which physical quantities such as phonon and electron dispersions repeat themselves (Kittel, 1986). It is constructed by drawing vectors from one reciprocal lattice point to another and then constructing lines perpendicular to these vectors at their midpoints. The smallest enclosed volume is the Brillouin zone.

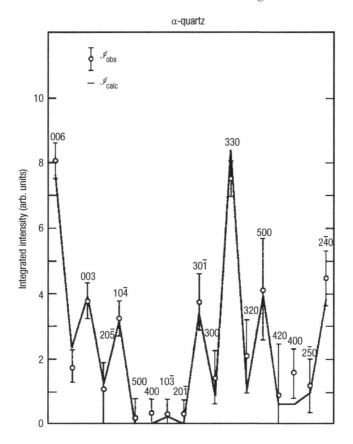

Fig. 2.3. Comparison of observed and calculated energy-integrated intensities for the soft mode in α-quartz at room temperature (from Axe and Shirane, 1970).

It is sometimes confusing as to which integral values of the Miller indices (*hkl*) correspond to proper reciprocal-lattice vectors. The latter are completely specified by the space group of the structure and may be determined from the "general conditions" limiting the (*hkl*) reflections listed under each Space Group heading in the *International Tables for Crystallography* (1995). It should be emphasized that the BZ centers and boundaries are determined *only* from these general conditions, and not the "special conditions" which accompany the special positions also listed under the Space Groups.

The intensity zone in reciprocal space is defined as the smallest unit over which the structure factor repeats itself. For the case where there is only one atom per unit cell, the BZ and IZ are identical; however, this is not generally the case when there is more than one atom per unit cell. We will demonstrate the distinction for the particular case of germanium. Ge has the diamond

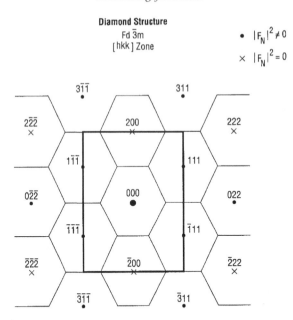

Fig. 2.4. Diagram of the (*hkk*) zone of reciprocal space for Ge. The thin lines outline Brillouin zones, while the thick line indicates an intensity zone. F_N is the static nuclear structure factor (see § 2.4.1).

structure (space group $Fd\bar{3}m$), consisting of two identical interpenetrating face-centered-cubic lattices with basis vectors $(0,0,0)$ and $(\frac{1}{4},\frac{1}{4},\frac{1}{4})$ (the 8*a* positions using Wyckoff notation) within the cubic unit cell. The general conditions limiting possible reflections are the standard fcc conditions with the Miller indices (*hkl*) all being even or odd. Because of the glide plane there are additional conditions governing the possible reflections which are $h + k + l = 2n + 1$ or $4n$, where n is an integer.

Figure 2.4 shows the (*hkk*) plane of the reciprocal lattice. Each reciprocal-lattice point is surrounded by its Brillouin zone, which has the shape of a hexagon. The fundamental properties of the solid, such as the electronic structure or the phonon frequencies, repeat with this periodicity. This is a very important condition, based on the translational invariance of the solid, which can be used to check the reality of observed features in experimental phonon spectra. If an inelastic peak is real, its energy should repeat from one Brillouin zone to the next. The intensity, however, can vary from zone to zone, and the repeat period is not always the same as the Brillouin zone. Because of the special conditions imposed by the glide plane in germanium, the repeat period for Bragg intensities (rectangular boundary in Fig. 2.4) is different from that of the Brillouin zone.

Another repeat distance in the reciprocal lattice is related to the periodicity of the dynamical structure factor for phonons [Eq. (2.34)]. This is more complicated than the static nuclear structure factor F_N and in order to calculate it, the atomic displacements associated with each mode of vibration must be known. Brockhouse and Iyengar (1958) showed for the case of germanium that the intensity zone for the phonons is the same as the intensity zone for Bragg scattering which is the large rectangle in Fig. 2.4. It is worthwhile to note that if the atoms are located at general positions which are not simple fractions of the unit-cell size, then there is no simple intensity zone since both the static and dynamical structure factors never repeat themselves.

2.5 Incoherent nuclear scattering

The differential cross section for incoherent scattering from a system consisting of a single element can be written as

$$\frac{d^2\sigma}{d\Omega_f dE_f}\bigg|_{\text{inc}} = \frac{k_f}{k_i}\frac{\sigma_{\text{inc}}}{4\pi}\frac{1}{2\pi\hbar}\sum_l \int_{-\infty}^{\infty} dt\, \left\langle e^{-i\mathbf{Q}\cdot\mathbf{r}_l(0)}e^{i\mathbf{Q}\cdot\mathbf{r}_l(t)}\right\rangle e^{-i\omega t}. \tag{2.52}$$

Note that the correlation function involved here is that of one atom with itself at different times. For elastic incoherent scattering from a crystal, this becomes

$$\frac{d\sigma}{d\Omega}\bigg|_{\text{inc}}^{\text{el}} = \frac{N}{4\pi}\sum_j \sigma_{\text{inc},j}e^{-2W_j}, \tag{2.53}$$

where N is the number of unit cells, the sum is over sites in a unit cell, and we have allowed for more than one type of element by including the site index j on the atomic incoherent cross section. As one can see, the $\mathbf{Q}$ dependence of the cross section is given entirely by the Debye–Waller factors. In general, incoherent scattering is roughly isotropic (at least for small Q and low temperatures), and it decreases monotonically with Q and T. The incoherent scattering from vanadium metal (for which $\sigma_{\text{inc}}/\sigma_{\text{coh}} = 276$) is commonly used as a standard for calibrating neutron flux at the sample position of a spectrometer.

The differential cross section for inelastic incoherent scattering from single phonons in a Bravais crystal is given by

$$\frac{d^2\sigma}{d\Omega_f dE_f}\bigg|_{\text{inc}} = \frac{\sigma_{\text{inc}}}{4\pi}\frac{k_f}{k_i}\frac{3N}{2m}\frac{e^{-2W}}{1-e^{-\hbar\omega/k_B T}}\frac{\langle(\mathbf{Q}\cdot\boldsymbol{\xi}_s)^2\rangle}{\omega}G(\omega), \tag{2.54}$$

where $G(\omega)$ is the density of states, and $\langle(\mathbf{Q}\cdot\boldsymbol{\xi}_s)^2\rangle$ is the average of $(\mathbf{Q}\cdot\boldsymbol{\xi}_s)^2$

over all modes with frequency ω. For a cubic crystal,

$$\langle (\mathbf{Q} \cdot \boldsymbol{\xi}_s)^2 \rangle = \tfrac{1}{3} Q^2. \tag{2.55}$$

For a non-Bravais crystal, the differential cross section becomes

$$\left. \frac{d^2\sigma}{d\Omega_f dE_f} \right|_{\text{inc}} = \frac{k_f}{k_i} \sum_j \frac{\sigma_{\text{inc},j}}{2m_j} \frac{e^{-2W}}{1 - e^{-\hbar\omega/k_B T}} \sum_s \frac{|\mathbf{Q} \cdot \boldsymbol{\xi}_{js}|^2}{\omega_s} \delta(\omega - \omega_s). \tag{2.56}$$

There are also multiphonon contributions to the incoherent scattering, which are less convenient to write down, but which are nevertheless present in a measurement.

Incoherent nuclear scattering is frequently an annoying background that one would like to eliminate. When the incoherence is due to a mixture of isotopes, use of a sample prepared with a single isotope can help to alleviate the problem. For incoherent scattering due to a finite nuclear spin, we will see in Chap. 8 that it is possible to eliminate that contribution with appropriate use of polarized neutrons. On the other hand, there are some cases where incoherent scattering is useful. In particular, the large incoherent cross section for ^{1}H can be used to study the dynamics of hydrogen atoms and molecules in solids. Examples are the study of hydrogen interstitials in Nb metal (Richter and Shapiro, 1980) and interstitial H_2 in solid C_{60} (FitzGerald *et al.*, 1999).

2.6 Coherent magnetic scattering

The neutron has a magnetic dipole moment equal to $-\gamma\mu_N\boldsymbol{\sigma}$, where γ ($= 1.913$) is the gyromagnetic ratio, μ_N is the nuclear magneton, and $\boldsymbol{\sigma}$ is the spin operator. The neutron can scatter from the magnetic moment of an atom via the dipole–dipole interaction. For simplicity, we will assume initially that the atomic moment is due purely to spin; later we will explain how the orbital moment can be taken into account. For an atom with net spin amplitude S, the magnetic moment in Bohr magnetons is gS, where g is the Landé splitting factor (with $g = 2$ for a spin-only moment). The amplitude for magnetic scattering can be written as pS, where

$$p = \left(\frac{\gamma r_0}{2} \right) g f(\mathbf{Q}), \tag{2.57}$$

and

$$\frac{\gamma r_0}{2} = 0.2695 \times 10^{-12} \text{ cm}, \tag{2.58}$$

with $r_0 = e^2/m_e c^2$ being the classical electron radius; e and m_e are the charge and mass of an electron, respectively, and c is the velocity of light.

The magnetic form factor $f(\mathbf{Q})$ is the Fourier transform of the normalized unpaired spin density $\rho_s(\mathbf{r})$ on an atom,

$$f(\mathbf{Q}) = \int \rho_s(\mathbf{r}) e^{i\mathbf{Q}\cdot\mathbf{r}} d\mathbf{r}, \tag{2.59}$$

with

$$f(0) \equiv 1. \tag{2.60}$$

The cross section for magnetic scattering was first derived by Halpern and Johnson (1939). It depends not only on the initial and final wave vectors of the neutron, but also on the corresponding neutron spin states, s_i and s_f, with $\mathbf{s} = \boldsymbol{\sigma}/2$. Generalizing Eq. (2.6) to take the neutron spin state and the magnetic interaction into account, the differential cross section can be written

$$\left.\frac{d^2\sigma}{d\Omega_f dE_f}\right|_{s_i s_f} = \frac{k_f}{k_i} \sum_{i,f} P(\lambda_i) \left| \left\langle \lambda_f \left| \sum_l e^{i\mathbf{Q}\cdot\mathbf{r}_l} U_l^{s_i s_f} \right| \lambda_i \right\rangle \right|^2 \delta(\hbar\omega + E_i - E_f). \tag{2.61}$$

The quantity $U_l^{s_i s_f}$ is the atomic scattering amplitude from the spin state s_i to s_f for atomic site l,

$$U_l^{s_i s_f} = \langle s_f | b_l - p_l \mathbf{S}_{\perp l} \cdot \boldsymbol{\sigma} + B_l \mathbf{I}_l \cdot \boldsymbol{\sigma} | s_i \rangle, \tag{2.62}$$

where b is the nuclear coherent scattering amplitude, B is the spin-dependent nuclear amplitude, and $\mathbf{I}$ is the nuclear spin operator. The quantity $\mathbf{S}_\perp$ is the magnetic interaction vector, first introduced by de Gennes (1963) and later adopted by Moon, Riste, and Koehler (1969),

$$\begin{aligned} \mathbf{S}_\perp &= \hat{\mathbf{Q}} \times (\mathbf{S} \times \hat{\mathbf{Q}}), \\ &= \mathbf{S} - \hat{\mathbf{Q}}(\hat{\mathbf{Q}} \cdot \mathbf{S}), \end{aligned} \tag{2.63}$$

where $\hat{\mathbf{Q}}$ is a unit vector along $\mathbf{Q}$. The notation expresses the fact that only the component of $\mathbf{S}$ perpendicular to $\mathbf{Q}$ contributes to the scattering amplitude. We note that

$$|\mathbf{S}_\perp|^2 = \sum_{\alpha,\beta} (\delta_{\alpha\beta} - \hat{Q}_\alpha \hat{Q}_\beta) S_\alpha^* S_\beta. \tag{2.64}$$

In this chapter we will limit ourselves to the consideration of scattering with unpolarized neutrons; applications involving polarized neutron beams will be treated in Chap. 8. For unpolarized neutrons scattering from a system containing a single species of magnetic atom, the differential cross section for atomic magnetic scattering can be written

$$\frac{d^2\sigma}{d\Omega_f dE_f} = \frac{N}{\hbar} \frac{k_f}{k_i} p^2 e^{-2W} \sum_{\alpha,\beta} (\delta_{\alpha,\beta} - \hat{Q}_\alpha \hat{Q}_\beta) S^{\alpha\beta}(\mathbf{Q}, \omega), \qquad (2.65)$$

with

$$S^{\alpha\beta}(\mathbf{Q}, \omega) = \frac{1}{2\pi} \int_{-\infty}^{\infty} dt \, e^{-i\omega t} \sum_l e^{i\mathbf{Q}\cdot\mathbf{r}_l} \langle S_0^\alpha(0) S_l^\beta(t) \rangle, \qquad (2.66)$$

where the angle brackets, $\langle\ldots\rangle$, denote an average over configurations. In deriving these formulas from Eq. (2.61), one also obtains a differential cross section for magneto-vibrational scattering, in which, for example, one excites phonons by scattering from ordered magnetic moments. For a discussion of magneto-vibrational scattering, see Lovesey (1984) or Squires (1978).

Integrating $S^{\alpha\beta}(\mathbf{Q}, \omega)$ over all frequencies, one obtains the Fourier transform of the instantaneous correlation function,

$$S^{\alpha\beta}(\mathbf{Q}, t = 0) = \int_{-\infty}^{\infty} d\omega \, S^{\alpha\beta}(\mathbf{Q}, \omega). \qquad (2.67)$$

If one also integrates over a Brillouin zone (BZ) in reciprocal space, a simple sum rule is obtained:

$$\int_{-\infty}^{\infty} d\omega \int_{BZ} d\mathbf{Q} \, S^{\alpha\beta}(\mathbf{Q}, \omega) = \frac{(2\pi)^3}{3v_0} S(S+1)\delta_{\alpha\beta}. \qquad (2.68)$$

This result is useful for a qualitative discussion of the distribution of magnetic scattering between elastic and inelastic processes. As we will discuss below, the elastic Bragg scattering from a magnetically ordered system is proportional to $\langle S^z \rangle^2$, where z is the axis along which the moments are ordered. For classical spins (large S), $\langle S^z \rangle = S$ at temperatures low compared to the ordering transition. The amount of sum rule weight that is available for spin-wave excitations is then proportional to S. The relative weight of magnon scattering compared to that in the magnetic Bragg peaks is then equal to $1/S$, which tends to zero in the limit of large S. On raising the temperature above the ordering transition, the scattering becomes diffuse, with a weight proportional to $S(S+1)$.

2.6.1 Paramagnetic scattering

An ideal paramagnet consists of a collection of completely independent, uncorrelated magnetic moments. In zero magnetic field, there is no energy barrier for spin fluctuations, so the scattering is completely elastic. For a system containing a single species of magnetic ions the differential scattering cross section is

$$\frac{d\sigma}{d\Omega_f} = \tfrac{2}{3}Np^2e^{-2W}S(S+1). \tag{2.69}$$

The scattering is Q-dependent due to the magnetic form factor (included in p) and the Debye–Waller factor.

This formula is inappropriate for analyzing scattering from the "paramagnetic" phase of a system with significant exchange coupling between the magnetic moments. In such a system, the scattering may be largely inelastic, and one must analyze it in terms of a generalized susceptibility, as discussed later in this chapter.

2.6.2 Magnetic form factors and orbital moments

Neutrons scatter from the magnetization density of an atom. In our discussion so far we have assumed that the magnetization density comes entirely from spin angular momentum; however, in general one must also allow for contributions from orbital angular momentum.

It is the Fourier transform of the magnetization density that enters the scattering cross section. For a spin-only moment we have

$$\mathbf{M}(\mathbf{Q})/\mu_B = gSf(\mathbf{Q}), \tag{2.70}$$

with $g = 2$. If $\Phi(r)$ is the radial wave function corresponding to the unpaired spin, then

$$f(\mathbf{Q}) = \int_0^\infty dr\, r^2 j_0(Qr)|\Phi(r)|^2 \equiv \langle j_0\rangle, \tag{2.71}$$

where $j_n(Qr)$ is a spherical Bessel function of order n, and we have ignored aspherical effects. The orbital moment can be included in a simple form, provided that Q is much smaller than the inverse of the mean radius of the wave function for the unpaired electrons (Lovesey, 1984). In this case,

$$\mathbf{M}(\mathbf{Q})/\mu_B = 2\langle j_0\rangle\mathbf{S} + (\langle j_0\rangle + \langle j_2\rangle)\mathbf{L}, \tag{2.72}$$

where $\mathbf{L}$ is the angular momentum vector.

For transition-metal ions, the spin density is generally associated with partially filled d orbitals. In a crystalline solid, the crystal field tends to break the degeneracy of the d orbitals, which commonly makes orbital currents (involving transitions between different d orbitals on the same site) energetically unfavorable. Nevertheless, some transition-metal ions exhibit small orbital moments, with the consequence that the Landé splitting factor

g differs slightly from its spin-only value of 2. In such a case, one can retain the parametrization given in Eq. (2.70), but with the form factor now given by

$$f(\mathbf{Q}) = \langle j_0 \rangle + \left(\frac{g-2}{2} \right) \langle j_2 \rangle. \tag{2.73}$$

For rare-earth ions, the orbital moment is generally unquenched. In this case, one must consider the total angular momentum $\mathbf{J}$ resulting from $\mathbf{S}$ and $\mathbf{L}$. Within the states of given $\mathbf{J}$ one has

$$2\mathbf{S} = g_S \mathbf{J}, \tag{2.74}$$

$$\mathbf{L} = g_L \mathbf{J}, \tag{2.75}$$

$$\mathbf{L} + 2\mathbf{S} = g \mathbf{J}, \tag{2.76}$$

where

$$g_S = \frac{J(J+1) - L(L+1) + S(S+1)}{J(J+1)}, \tag{2.77}$$

$$g_L = \frac{J(J+1) + L(L+1) - S(S+1)}{2J(J+1)}, \tag{2.78}$$

and $g = g_S + g_L$. In the scattering formulas, one should replace $g\mathbf{S}$ with $g\mathbf{J}$, and the form factor is given by

$$f(\mathbf{Q}) = \langle j_0 \rangle + \frac{g_L}{g} \langle j_2 \rangle. \tag{2.79}$$

Parametrized versions of the $\langle j_n \rangle$ needed to calculate form factors have been tabulated for magnetic ions by Brown (1995). Deviations of the spin density from spherical symmetry are often of negligible importance except at large Q. An important counter-example is Cu^{2+}, where the unpaired spin density involves a single 3d orbital, resulting in substantial anisotropy in $f(\mathbf{Q})$ (Shamoto *et al.*, 1993).

2.6.3 Determination of magnetic structures with unpolarized neutrons

The coherent elastic differential cross section for magnetic scattering from a magnetically ordered crystal is given by

$$\frac{d\sigma}{d\Omega_f}\bigg|_{coh}^{el} = N_M \frac{(2\pi)^3}{v_M} \sum_{\mathbf{G}_M} \delta(\mathbf{Q} - \mathbf{G}_M) |F_M(\mathbf{G}_M)|^2, \tag{2.80}$$

where F_M, the static magnetic structure factor, is given by

$$\mathbf{F}_M(\mathbf{G}_M) = \sum_j p_j \mathbf{S}_{\perp j} e^{i\mathbf{G}_M \cdot \mathbf{d}_j} e^{W_j}. \tag{2.81}$$

We have introduced the subscript M to indicate that v_M and N_M refer to the volume of the magnetic unit cell and the number of such cells in the sample, respectively; similarly, the sum in Eq. (2.81) is over sites within the magnetic unit cell. Except for the case of ferromagnetism, the magnetic unit cell is typically larger than the chemical unit cell. Correspondingly, a larger magnetic unit cell will lead to new reciprocal-lattice vectors, $\mathbf{G}_M$. The relationship between the integrated intensity of a magnetic Bragg peak and $|\mathbf{F}_M(h_M k_M l_M)|$ is the same as that for nuclear scattering. One can make use of Eqs. (2.27) and (2.29) substituting $|\mathbf{F}_M|$ for F_N.

In determining a magnetic structure, one usually knows the crystal structure in advance (i.e., the atomic positions $\mathbf{d}_j$ within a unit cell are known). The parameters to be determined, then, are the amplitudes and orientations of the spins, which can be deduced from a measured set of $|\mathbf{F}_M|$. We do not intend to discuss a general scheme of solving magnetic structures here. As reviewed by Cox (1972), sophisticated techniques exist for solving the spin arrangement given a unique set of $|\mathbf{F}_M(hkl)|$ (where we will now drop the M subscript on the Miller indices for simplicity). Instead, we will focus on some of the ambiguities that can occur when one analyzes measurements on a real (as opposed to ideal) sample.

Given the vector nature of the magnetic moments, there is generally more than one way to align a magnetic structure with respect to the chemical lattice. All things being equal, each possible orientation of the magnetic structure will occur as a distinct domain within a crystal, with the volume of the crystal distributed equally among the various domains. For example, in ferromagnetic iron, the spins can choose to align along any of the three $\langle 100 \rangle$ axes, with two possible orientations along each axis. For a sample cooled in zero magnetic field, the Fe-spins will be equally distributed among the six domains, regardless of whether the sample is a single crystal or powder. An extra complication that can occur in powder measurements is the overlap of non-equivalent reflections, such as cubic (300) and (221). In a tetragonal crystal, it is possible for the $(hk0)$ and $(h\bar{k}0)$ to have different magnetic structure factors; obviously, this distinction is lost when domains are present.

In principle, magneto-elastic couplings can cause structural distortions that would allow a distinction between otherwise equivalent domains. In iron metal, for example, it is obvious that the [100] spin direction in the ferromagnetic state breaks cubic symmetry. The unit cell must be tetragonally distorted; however, no evidence of such a crystallographic distortion has been observed. There are some systems, such as CoO and several rare-earth elements and rare-earth pnictides, in which a "giant" magneto-elastic coupling is present, yielding a distortion reflecting the magnetic configurations. In such cases, a single crystal with a specially chosen shape may give rise to an unbalanced domain distribution. More commonly, the magneto-elastic distortions are too small to be of use. Sometimes, the anisotropy found in a measurement such as the magnetic susceptibility of a single crystal may give crucial information on the spin directions; frequently, however, one must solve the magnetic structure without knowing the true magnetic symmetry of the crystal under examination.

Another frequent difficulty in determining magnetic structures is that it is possible for different magnetic structures to yield identical Bragg peak intensities. This problem is rare in nuclear structural determinations [although we note the special case of soft-mode condensation, where the degenerate soft modes create similar problems, e.g., the M_3 rotation in a perovskite as reported by Sato *et al.* (1982)]. In contrast, the problem of ambiguities in magnetic order is more fundamental. To see this, we first note that magnetic structures can be classified into two categories, *collinear* and *noncollinear*. In the collinear case, all magnetic spins are aligned along the same axis, either parallel or antiparallel. In some cases, a noncollinear structure yields the same Bragg intensities as a collinear model. Three examples are shown in Fig. 2.5 based on tetragonal chemical structures. Examples (I) and (II) are fictitious models for demonstration. (III) is actually observed in K_2NiF_4 and La_2CuO_4. In each case, Bragg intensities from equal populations of domains (a) and (b) are equal to those from structure (c). Note that I(c) is a collinear structure, while II(c) and III(c) are noncollinear.

For collinear structures, one can work out explicit symmetry-dependent formulas for domain-averaged magnetic intensities (Shirane, 1959). Figure 2.6 shows some examples of magnetic structures with different structural and magnetic symmetries. To analyze the effects of averaging over domains, it is convenient to factorize the magnetic structure factor as follows:

$$\mathbf{F}_M = \mathbf{S}_\perp \tilde{F}_M, \tag{2.82}$$

where $\mathbf{S}_\perp$ represents the component of the spin axis perpendicular to $\mathbf{Q}$, and

$$\tilde{F}_M = \sum_j p_j e^{-W_j} e^{i\mathbf{Q}\cdot\mathbf{d}_j}, \tag{2.83}$$

with p_j now set to $+|p_j|$ or $-|p_j|$, depending on the relative spin direction. Note that both $\mathbf{S}_\perp$ and $\tilde{F}_M$ depend on (hkl). The set of p_js defines the "configurational" symmetry of the magnetic structure, which need not be the same as the "chemical" symmetry. The symmetries of the structures shown in Fig. 2.6 are listed in Table 2.1.

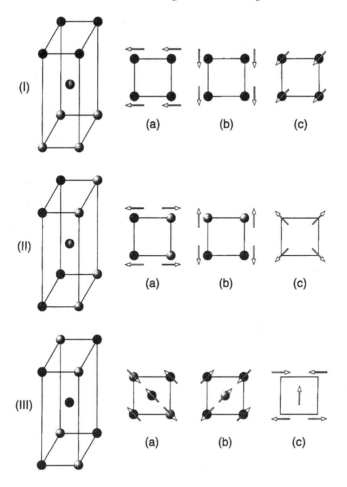

Fig. 2.5. Three examples of magnetic structures in a tetragonal lattice. The domain-averaged Bragg intensities from (a) + (b) are identical to those given by (c).

The configurational symmetry allows several $\tilde{F}_M$ with different combinations of $\pm h, \pm k, \pm l$ to have the same absolute value, although each (hkl) may yield a distinct value of $S_\perp^2$. For example, (hkl), (klh), and (lhk) planes are equivalent if the symmetry is rhombohedral. This group of equivalent hkl reflections is usually represented as $\{hkl\}$. When we average over domains, we find

$$\langle |\mathbf{F}_M(\{hkl\})|^2 \rangle = \langle |\mathbf{S}_\perp|^2 \rangle |\tilde{F}_M(\{hkl\})|^2. \tag{2.84}$$

If we label the angle between $\mathbf{S}$ and $\mathbf{Q}$ as η, then

$$\langle |\mathbf{S}_\perp|^2 \rangle = S^2(1 - \langle \cos^2 \eta \rangle). \tag{2.85}$$

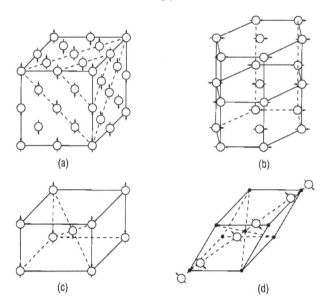

Fig. 2.6. Examples of antiferromagnetic spin arrangements: (a) MnO-type, (b) NiAs-type, (c) MnF_2-type (Rutile-type), (d) Fe_2O_3-type (from Shull and Wollan, 1956).

Thus, the problem of averaging over domains is reduced to evaluating $\langle \cos^2 \eta \rangle$ for each crystallographic system.

To obtain a general formula for $\cos \eta$, we note that the scattering vector $\mathbf{G}_M$ is perpendicular to the plane (hkl), and we describe the plane normal to the spin direction by $(h_0 k_0 l_0)$. We can then make use of the formula for the cosine of the angle between the vectors normal to (hkl) and $(h_0 k_0 l_0)$ giving, (James, 1954)

$$\cos \eta = \left(h\mathbf{a}^* + k\mathbf{b}^* + l\mathbf{c}^* \right) \cdot \left(h_0\mathbf{a}^* + k_0\mathbf{b}^* + l_0\mathbf{c}^* \right) \frac{dd_0}{4\pi^2}, \qquad (2.86)$$

where $\mathbf{a}^*$, $\mathbf{b}^*$, and $\mathbf{c}^*$ are the primitive lattice vectors in reciprocal space, and $d(hkl)$ and $d_0(hkl)$ are the spacings of the planes (hkl) and $(h_0 k_0 l_0)$ in real space. To evaluate $\langle \cos^2 \eta \rangle$, one must square Eq. (2.86) and average it over equivalent (hkl). A table of equivalent reflections in various systems is given in *International Tables for Crystallography* (1995).

The direction of the spin can be specified in terms of the angles φ_a, φ_b, and φ_c between $\mathbf{S}$ and the crystallographic axes $\mathbf{a}$, $\mathbf{b}$, and $\mathbf{c}$. The cosines of these angles can be expressed in terms of h_0, k_0, and l_0 by making use of Eq. (2.86), with appropriate choices for h, k, and l. As examples, we will calculate $\langle \cos^2 \eta \rangle$ for cubic and tetragonal symmetries, and then list the results for other cases.

Table 2.1. *Crystal symmetry and configurational symmetry of magnetic structures shown in Fig. 2.6. The abbreviation c.p. stands for cyclic permutations of coordinates (Shirane, 1959).*

Spin arrangement	Crystal structure	Magnetic structure
(a) MnO-type	Cubic, $Fm3m$ NaCl-type with a_0	Rhombohedral, $R\bar{3}, a = 2a_0$ $+p$ at $(0,0,0), (0, \frac{1}{2}, \frac{1}{2}) +$ c.p., $\pm(0, \frac{1}{4}, \frac{3}{4}) +$ c.p., $\pm(\frac{1}{4}, \frac{1}{4}, \frac{1}{2}) +$ c.p.
(b) NiAs-type	Hexagonal, $P6_3mc$ Ni at $(0,0,0), (0,0,\frac{1}{2})$	Hexagonal, $P6mm$ $+p$ at $(0,0,0)$ $-p$ at $(0,0,\frac{1}{2})$
(c) MnF$_2$-type (Rutile-type)	Tetragonal, $P4_2mmm$ Mn at $(0,0,0), (\frac{1}{2},\frac{1}{2},\frac{1}{2})$	Tetragonal, $P4/mm$ $+p$ at $(0,0,0)$ $-p$ at $(\frac{1}{2},\frac{1}{2},\frac{1}{2})$
(d) Fe$_2$O$_3$-type	Rhombohedral, $R\bar{3}c$ Fe at $\pm(u,u,u)$, $\pm(\frac{1}{2}+u, \frac{1}{2}+u, \frac{1}{2}+u)$	Rhombohedral, $R3c$ for Cr$_2$O$_3$, $R\bar{3}$ for Fe$_2$O$_3$ $+p$ at (u,u,u), $(\frac{1}{2}+u, \frac{1}{2}+u, \frac{1}{2}+u)$ $-p$ at $(-u,-u,-u)$, $(\frac{1}{2}-u, \frac{1}{2}-u, \frac{1}{2}-u)$

Note: a_0 is the lattice parameter of the nuclear unit cell; the lattice parameter, a, of the magnetic unit cell is $2a_0$

(A) Cubic

$$\cos\eta = (hh_0 + kk_0 + ll_0) a^{*2} \frac{dd_0}{4\pi^2}, \tag{2.87}$$

$$\begin{aligned} \langle\cos^2\eta\rangle &= \tfrac{1}{3}\left(h^2 + k^2 + l^2\right)\left(h_0^2 + k_0^2 + l_0^2\right) a^{*4} \frac{dd_0}{16\pi^4} \\ &= \tfrac{1}{3}\left(\cos^2\varphi_a + \cos^2\varphi_b + \cos^2\varphi_c\right) \\ &= \tfrac{1}{3}. \end{aligned} \tag{2.88}$$

(B) Tetragonal

$$\cos\eta = \left[(hh_0 + kk_0) a^{*2} + ll_0 c^{*2}\right] \frac{dd_0}{4\pi^2}. \tag{2.89}$$

By summing up the terms for equivalent reflections (hkl), $(hk\bar{l})$, $(k\bar{h}l)$, and $(k\bar{h}\bar{l})$, we obtain

$$\begin{aligned} \langle\cos^2\eta\rangle &= \left[\tfrac{1}{2}\left(h^2 + k^2\right)\left(h_0^2 + k_0^2\right) a^{*4} + l^2 l_0^2 c^{*4}\right] \frac{dd_0}{16\pi^4} \\ &= \left[\tfrac{1}{2}\left(h^2 + k^2\right) a^{*2} \sin^2\varphi_c + l^2 c^{*2} \cos^2\varphi_c\right] \frac{d}{4\pi^2}. \end{aligned} \tag{2.90}$$

For the tetragonal case, we see that only one of the three direction cosines can be determined, while for cubic symmetry domain-averaging eliminates all information about the spin direction. As shown by Shirane (1959), one is also limited to a single direction cosine for the hexagonal and rhombohedral cases. The orthorhombic case is the only one for which information on all three angles survives.

(C) *Hexagonal*

$$\langle \cos^2 \eta \rangle = \left[\tfrac{1}{2} \left(h^2 + k^2 + hk \right) a^{*2} \sin^2 \varphi_c + l^2 c^{*2} \cos^2 \varphi_c \right] \frac{d^2}{4\pi^2}. \tag{2.91}$$

(D) *Rhombohedral*

$$\langle \cos^2 \eta \rangle = \tfrac{1}{3} \left[(n - r) \left(1 - \cos \alpha^* \right) \sin^2 \varphi \right.$$
$$\left. + (n + 2r) \left(1 + 2 \cos \alpha^* \right) \cos^2 \varphi \right] a^{*2} \frac{d^2}{4\pi^2}, \tag{2.92}$$

with

$$n = h^2 + k^2 + l^2, \quad r = hk + kl + lh,$$

where φ is the angle between the spin direction and the [111] axis, and $\alpha*$ is the angle between each pair of primitive reciprocal-lattice vectors.

(E) *Orthorhombic*

$$\langle \cos^2 \eta \rangle = \left(h^2 a^{*2} \cos^2 \varphi_a + k^2 b^{*2} \cos^2 \varphi_b + l^2 c^{*2} \cos^2 \varphi_c \right) \frac{d^2}{4\pi^2}. \tag{2.93}$$

2.7 Coherent inelastic magnetic scattering

2.7.1 Magnons

Thermal energy and quantum zero-point fluctuations cause the relative orientation of individual magnetic moments in an ordered structure to fluctuate. Because the spins are coupled to one another by exchange interactions, the normal modes of these fluctuations are collective excitations called spin waves. The energy of the spin waves, $\hbar\omega$, exhibits dispersion with respect to the wave vector $\mathbf{q}$ within a Brillouin zone. As the energy of a spin wave is quantized, it is often referred to as a magnon. A neutron, scattering from a magnetic system, can absorb or emit one or more magnons; here we will consider only the single-magnon cross section.

The appropriate starting point for a system containing a single species of magnetic ion is the double-differential cross section given by Eq. (2.65). We begin by noting that spin waves involve displacements of the spins in the directions perpendicular to the average spin direction. It is convenient to choose the average spin direction to be along z; it follows that spin waves involve S^x and S^y. A proper analysis (Lovesey, 1984; Squires, 1978) then leads to the result

$$\sum_{\alpha,\beta}(\delta_{\alpha,\beta} - \hat{Q}_\alpha\hat{Q}_\beta)S^{\alpha\beta}(\mathbf{Q},\omega) = \tfrac{1}{2}(1 + \hat{Q}_z^2)S_{sw}(\mathbf{Q},\omega). \tag{2.94}$$

If magnetic domains are present, one must evaluate the appropriate average of $\hat{Q}_z^2$. (For example, for ferromagnetic domains in a cubic crystal, $\langle\hat{Q}_z^2\rangle = \tfrac{1}{3}$.) To write down the inelastic scattering function for spin waves, $S_{sw}(\mathbf{Q},\omega)$, we must differentiate between the cases of ferromagnets and antiferromagnets.

2.7.1.1 Ferromagnet

For a Heisenberg ferromagnet with only nearest-neighbor interactions, the spin-wave dispersion at small $\mathbf{q}$ is given by

$$\hbar\omega_\mathbf{q} = Dq^2, \tag{2.95}$$

with

$$D = 2JSa^2, \tag{2.96}$$

where J is the exchange energy. The inelastic scattering function can be written as

$$\begin{aligned}
S_{sw}(\mathbf{Q},\omega) = S\sum_{\mathbf{G},\mathbf{q}}\bigl[&(n_\mathbf{q}+1)\delta(\mathbf{Q}-\mathbf{q}-\mathbf{G})\delta(\omega-\omega_\mathbf{q}) \\
&+ n_\mathbf{q}\,\delta(\mathbf{Q}+\mathbf{q}-\mathbf{G})\delta(\omega+\omega_\mathbf{q})\bigr].
\end{aligned} \tag{2.97}$$

Interactions between magnons lead to damping. To include the effect of damping it is convenient to make use of the fluctuation–dissipation theorem, Eq. (2.31); the appropriate form of the imaginary part of the dynamic susceptibility is then

$$\chi''(\mathbf{Q},\omega) = S\left(\frac{\omega_\mathbf{q}}{\omega_\mathbf{q}'}\right)\left[\frac{\Gamma}{(\omega-\omega_\mathbf{q}')^2+\Gamma^2} - \frac{\Gamma}{(\omega+\omega_\mathbf{q}')^2+\Gamma^2}\right], \tag{2.98}$$

where

$$\omega_\mathbf{q}'^2 = \omega_\mathbf{q}^2 - \Gamma^2. \tag{2.99}$$

As an example, Fig. 2.7 shows the observed spectra for spin waves measured in the ferromagnetic alloy $Fe_{0.34}Cr_{0.66}$ at several temperatures below the Curie temperature T_C (Shapiro et al., 1981). The peak at $E = 0$ is due to incoherent elastic scattering; the magnon peaks are at $E = \pm 0.4\,\text{meV}$. One

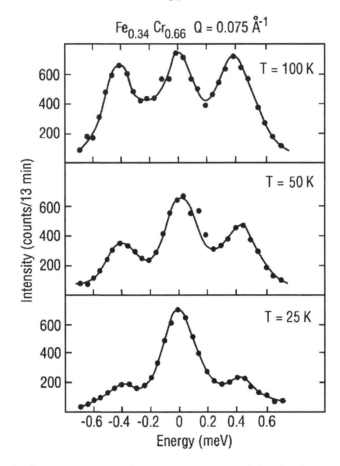

Fig. 2.7. Inelastic neutron scattering spectra measured in the ferromagnetic alloy $Fe_{0.34}Cr_{0.66}$ ($T_C = 330\,K$) at $Q = 0.075\,\text{Å}^{-1}$, near the forward direction. The two peaks at finite energy correspond to creation of a magnon ($+0.4\,meV$) and annihilation of a magnon ($-0.4\,meV$). The peak at $E = 0\,meV$ is mostly due to elastic incoherent scattering (from Shapiro *et al.*, 1981).

can see that the intensity is proportional to T, as expected from the Bose factor in the limit $\hbar\omega \ll k_B T$.

2.7.1.2 Antiferromagnet

In an antiferromagnet, the spin-wave intensity is strong near an antiferromagnetic superlattice peak and weak near a Bragg peak of the chemical lattice. For a Heisenberg antiferromagnet with only nearest-neighbor interactions, the dispersion at small q is given by

$$\hbar\omega_{\mathbf{q}} = \hbar c q, \tag{2.100}$$

where

$$c = zJSa/\hbar \tag{2.101}$$

is the spin-wave velocity, z is the number of nearest neighbors, and J is the superexchange energy. For small q measured with respect to antiferromagnetic superlattice vectors $\mathbf{G_M}$, the spin-wave scattering function can be written as

$$
\begin{aligned}
S_{sw}(\mathbf{Q}, \omega) = S \sum_{\mathbf{G_M, q}} \frac{\hbar\omega_0}{\hbar\omega_q} & [(n_q + 1)\delta(\mathbf{Q} - \mathbf{q} - \mathbf{G_M})\delta(\omega - \omega_q) \\
& + n_q \delta(\mathbf{Q} + \mathbf{q} - \mathbf{G}_m)\delta(\omega + \omega_q)],
\end{aligned}
\tag{2.102}
$$

where

$$\hbar\omega_0 = 2zJS. \tag{2.103}$$

The factor $1/\omega_q$ makes the scattering function different from that for a ferromagnet, but similar to the scattering function for phonons. When damping of antiferromagnetic spin waves is important, it can be included using Eq. (2.45).

2.7.2 Diffuse magnetic scattering

In analyzing diffuse magnetic scattering, it is frequently convenient to discuss the scattering in terms of the dynamic susceptibility, rather than $S(\mathbf{Q}, \omega)$. Before continuing, we will first consider some of the properties of generalized susceptibility.

The generalized susceptibility $\chi(\mathbf{Q}, \omega)$ is a complex function with real and imaginary parts, χ' and χ'', respectively. As a result of its connection with correlation functions, $\chi''(\mathbf{Q}, \omega)$ must be an odd function of frequency. The real and imaginary parts of the susceptibilty are related by the Kramers–Kronig relation:

$$\chi'(\mathbf{Q}, \omega) = \frac{1}{\pi} \int_{-\infty}^{\infty} d\omega' \frac{\chi''(\mathbf{Q}, \omega')}{\omega - \omega'}. \tag{2.104}$$

For the static susceptibility, one has

$$\chi'(\mathbf{Q}, 0) = -\frac{1}{\pi} \int_{-\infty}^{\infty} d\omega' \frac{\chi''(\mathbf{Q}, \omega')}{\omega'}. \tag{2.105}$$

In the limit $\mathbf{Q} \to 0$, $\chi'(\mathbf{Q}, 0)$ becomes the bulk (uniform) susceptibility, χ_b,

$$\chi'(\mathbf{0}, 0) = \chi_b. \tag{2.106}$$

Thus, Eqs. (2.105) and (2.106) provide a connection between the imaginary part of the susceptibility, measured by neutron scattering, and the bulk susceptibility, determined with a dc magnetometer or a magnetic resonance technique. It is possible to choose a general form for χ'' which automatically satisfies the relations above:

$$\chi''(\mathbf{Q}, \omega) = \chi'(\mathbf{Q}, 0) \, \omega \, F(\mathbf{Q}, \omega), \tag{2.107}$$

where the spectral weight function $F(\mathbf{Q}, \omega)$ is an even function of ω satisfying the normalization condition

$$\int_{-\infty}^{\infty} F(\mathbf{Q}, \omega) \, d\omega = 1. \tag{2.108}$$

2.7.2.1 Quasielastic scattering

In some systems, such as mixed-valence and heavy-fermion compounds, localized moments (e.g., spin and orbital moments from 4f electrons in rare-earth ions) interact with delocalized conduction electrons, but have only weak interactions with one another. In such cases, the lack of spatial correlations means that the susceptibility is essentially independent of $\mathbf{Q}$ (so that the magnetic scattering depends on $\mathbf{Q}$ mainly through the magnetic form factor.) If the spin excitations are damped exponentially in time, then an appropriate form for the spectral-weight function is a lorentzian,

$$F(\mathbf{Q}, \omega) = \frac{1}{\pi} \frac{\Gamma}{\omega^2 + \Gamma^2}, \tag{2.109}$$

so that

$$\chi''(\mathbf{Q}, \omega) = \frac{\chi'(\mathbf{0}, 0)}{\pi} \frac{\omega \Gamma}{\omega^2 + \Gamma^2}. \tag{2.110}$$

The adjective "quasielastic" comes from application of this model to situations where Γ is relatively small (i.e., comparable to the energy resolution); however, the usefulness of the model extends beyond such cases. At the limits of high and low temperature, the scattering function becomes

$$S(\mathbf{Q}, \omega) = \frac{\chi'(\mathbf{0}, 0)}{\pi} \frac{\Gamma}{\omega^2 + \Gamma^2} \times \begin{cases} k_B T & \text{for } k_B T \gg \hbar\omega, \\ \omega & \text{for } k_B T \ll \hbar\omega. \end{cases} \tag{2.111}$$

At high temperature, the scattering has a lorentzian frequency dependence, with a maximum at $\omega = 0$; however, at low temperature the extra factor of ω multiplying the lorentzian causes the intensity maximum to shift to $\omega_{max} = \Gamma$. An example of the temperature evolution of $S(\mathbf{Q}, \omega)$ (with a temperature-independent damping constant Γ) is shown in Fig. 2.8.

In some metallic rare-earth and actinide compounds, the interaction between the more localized f electrons and the conduction electrons can lead to a rather large and temperature-dependent Γ. Figure 2.9 shows $\chi''(\mathbf{Q}, \omega)$

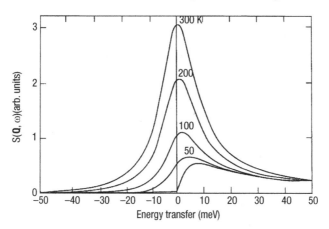

Fig. 2.8. Examples of quasielastic scattering at several temperatures (K) for $\hbar\Gamma = 8.6$ meV.

measured for the mixed-valence compound $Ce_{0.74}Th_{0.26}$ at several temperatures (Shapiro *et al.*, 1977). At high temperature, $\Gamma \approx 20$ meV and increases with decreasing T. Near the valence transition at $T_v \approx 150$ K, the linewidth changes abruptly to $\Gamma > 70$ meV. This change indicates a drastic reduction in the lifetime of spin excitations as the hybridization between the f electrons and the conduction electrons increases.

2.7.2.2 Critical scattering

Here we return to considering systems that order magnetically. Above the ordering temperature T_c, the spin fluctuations at small $\mathbf{q}$ (with $\mathbf{q} = \mathbf{Q} - \mathbf{G}_M$) are overdamped. The mean-field form for the wave-vector-dependent susceptibility at small $\mathbf{q}$ is

$$\chi'(\mathbf{Q}, 0) = \frac{S(S+1)}{r_1^2(\kappa^2 + q^2)}, \qquad (2.112)$$

where κ is equal to the inverse of the magnetic correlation length, r_1 is a measure of the range of the exchange interactions, and

$$(r_1\kappa)^2 = (T - T_c)/T_c. \qquad (2.113)$$

For the spectral-weight function, the lorentzian form of Eq. (2.109) is appropriate, but with the change $\Gamma \to \Gamma_{\mathbf{q}}$. For a ferromagnet, diffusion theory gives

$$\Gamma_{\mathbf{q}} = \Lambda q^2, \qquad (2.114)$$

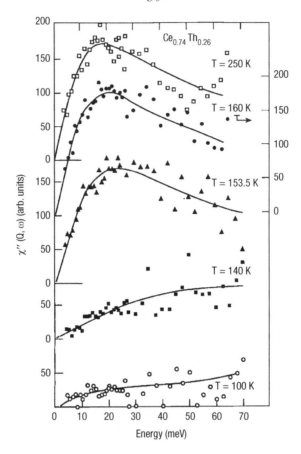

Fig. 2.9. Magnetic $\chi''(Q,\omega)$ measured in a polycrystalline sample of the mixed-valence compound $Ce_{0.74}Th_{0.26}$, which exhibits a large change in the linewidth, Γ, at the valence transition temperature $T_v = 150$ K. The low-energy range ($E < 27.5$ meV) was measured at $Q = 1.5$ Å^{-1}, while data at $E > 27.5$ meV were obtained at $Q = 3.0$ Å^{-1} (from Shapiro *et al.*, 1977).

where Λ is the spin diffusion constant, which varies with temperature as $\sqrt{\kappa}$. Dynamic scaling theory predicts that

$$\Gamma_{\mathbf{q}} = Bq^{5/2}f(\kappa/q),\qquad(2.115)$$

where B is a constant and $f(\kappa/q)$ is a scaling function. For $q \ll \kappa$, $f(\kappa/q)$ is proportional to $(\kappa/q)^{1/2}$, so that Eq. (2.115) becomes equivalent to Eq. (2.114).

Figure 2.10 shows inelastic spectra (obtained with polarized neutrons) for a single crystal of iron at 22 K above the Curie temperature, measured about the (1,1,0) zone center along the [110] direction (Wicksted, Böni, and Shirane 1984). As q increases from 0.05 to 0.15, the peak intensity decreases

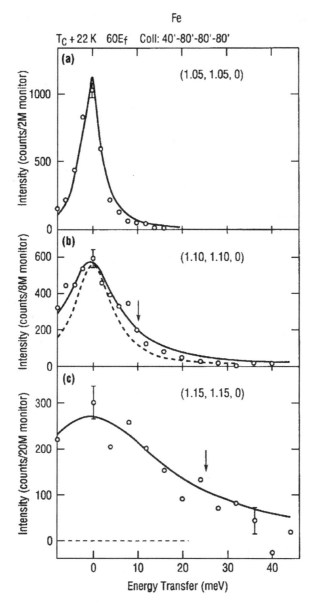

Fig. 2.10. Paramagnetic scattering spectra measured in Fe using polarized neutron beams. Measurements were performed at $T = 1.02T_C$ for several different **Q** values. The arrows point toward expected spin-wave peaks. The solid line is a fit with Eq. (2.116) with the resolution function; the dashed curve in (b) shows the normalized cross section without resolution correction (from Wicksted, Böni, and Shirane, 1984).

dramatically and the linewidth broadens. The solid curves are fits using

$$S(\mathbf{Q}, \omega) = \frac{2}{3} \frac{S(S+1)}{r_1^2} \frac{1}{\kappa^2 + q^2} \frac{1}{\pi} \frac{\Gamma}{\Gamma^2 + \omega^2} \frac{\hbar\omega/k_B T}{1 - e^{\hbar\omega/k_B T}}, \qquad (2.116)$$

with the values for κ and Γ determined by independent measurements.

References

Axe, J. D. and Shirane, G. (1970). *Phys. Rev. B* **1**, 342.

Bacon, G. E. (1975). *Neutron Diffraction* (Clarendon Press, Oxford).

Brockhouse, B. and Iyengar, P. K. (1958). *Phys. Rev.* **111**, 747.

Brown, P. J. (1995). In *International Tables for Crystallography*, Volume C, ed. A. J. C. Wilson (Kluwer Academic Publishers, Dordrecht) p. 391.

Chaikin, P. M. and Lubensky, T. C. (1995). *Principles of Condensed Matter Physics* (Cambridge University Press, Cambridge).

Cox, D. E. (1972). *IEEE Trans. Magn.* **MAG-8**, 161.

de Gennes, P. G. (1963). In *Magnetism*, ed. G. T. Rado and H. Suhl, (Academic Press, New York), Vol. III, p. 115.

FitzGerald, S. A., Yildirim, T., Santodonato, L. J., Neumann, D. A., Copley, J. R. D., Rush, J. J., and Trouw, F. (1999). *Phys. Rev. B* **60**, 6439.

Halpern, O. and Johnson, M. R. (1939). *Phys. Rev.* **55**, 898.

Harada, J., Axe, J. D., and Shirane, G. (1970). *Acta. Cryst. A* **26**, 608.

International Tables for Crystallography (1995). Volume A: *Space Group Symmetry*, ed. T. Hahn (D. Riedel Publishing Co., Boston, MA).

Ishikawa, Y., Noda, Y., Uemura, Y. J., Majkrzak, C. F., and Shirane, G. (1985). *Phys. Rev. B* **31**, 5884.

James, R. W. (1954). The Optical Principles of the Diffraction of X-Rays (Bell, London) p. 604.

Kittel, C. (1986). *Introduction to Solid State Physics* (John Wiley & Sons, New York).

Lovesey, S. W. (1984). *Theory of Neutron Scattering from Condensed Matter* (Clarendon Press, Oxford).

Marshall, W. and Lowde, R. D. (1968). *Rept. Progr. Phys.* **31**, 705.

Moon, R. M., Riste, T., and Koehler, W. C. (1969). *Phys. Rev.* **181**, 920.

Richter, D. and Shapiro, S. M. (1980). *Phys. Rev. B* **22**, 599.

Sato, M., Grier, B. H., Shirane, G., and Akahane, T. (1982). *Phys. Rev. B* **25**, 6876.

Shamoto, S., Sato, M., Tranquada, J. M., Sternlieb, B. J., and Shirane, G. (1993). *Phys. Rev. B* **48**, 13817.

Shapiro, S. M., Axe, J. D., Birgeneau, R. J., Lawrence, J. M., and Parks, R. D. (1977). *Phys. Rev. B* **16**, 2225.

Shapiro, S. M., Fincher, Jr, C. R., Palumbo, A. C., and Parks, R. D. (1981). *Phys. Rev. B* **24**, 6661.

Shirane, G. (1959). *Acta Cryst.* **12**, 282.

Shirane, G. (1974). *Rev. Mod. Phys.* **46**, 437.

Shull, C. G. and Wollan, E. O. (1956). *Solid State Physics* **2**, 137.

Squires, G. L. (1978). *Introduction to the Theory of Thermal Neutron Scattering* (Cambridge Press, New York).

Van Hove, L. (1954). *Phys. Rev.* **95**, 249.

Wicksted, J. P., Böni, P., and Shirane, G. (1984). *Phys. Rev. B* **30**, 3555.

3

Elements of a three-axis instrument

The three axes involved in a triple-axis instrument are the monochromator axis, the sample axis and the analyzer axis as shown schematically in Fig. 3.1. However, there are many other elements of a three-axis instrument which are necessary in order to have an efficient spectrometer. These include monochromator and analyzer crystals, energy filters, collimators, and detectors. Since each axis and crystal has to be individually moved for every setting of the instrument, computer control is essential. In addition, extensive shielding is required in order to protect the experimenter and to reduce the overall background or noise in the experiment. In this chapter we shall discuss each of these elements in some detail.

3.1 Shielding

Most modern reactors have beam tubes which do not look directly at the core, but instead have their axes tangential to it. This decreases the quantity of unwanted high-energy ($\gtrsim 200 \, \text{meV}$) neutrons (also called fast neutrons) in the beam tube, since only those neutrons that have a component of velocity nearly parallel to the beam-tube axis can enter the monochromator area. Neutrons which satisfy this condition have necessarily undergone collisions with the moderating material and have lost considerable energy compared to fission energies. The moderation is never perfect, and there are always some fast neutrons that enter the monochromator area, along with γ-rays and other unwanted radiation. It is therefore necessary to shield adequately against this unwanted radiation. The shielding should contain a combination of materials that can scatter the fast neutrons, slow them down, and then absorb them. In addition, a significant amount of lead is needed to absorb the dangerous γ-rays. Typical shields are made of large amounts of iron, in the form of steel balls to scatter the fast neutrons, large amounts of

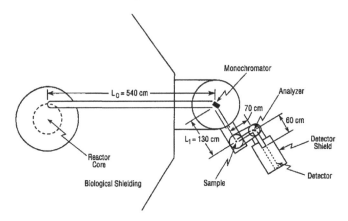

Fig. 3.1. Schematic of the three-axis instrument formerly situated at the H7 beam port of the HFBR at Brookhaven National Laboratory.

hydrogen in the form of concrete, resin, wax or polyurethane to slow them down, and either boron or cadmium to absorb the low-energy neutrons. A popular material used by most facilities is a boron epoxy which contains lots of hydrogen and boron. The advantage of this material is that it can be cast into any desired shape.

Many collisions are needed to slow the fast neutrons, so the shielding around a monochromator is bulky and usually consists of a large drum, approximately 2 m in diameter, filled with the mixture of scattering and absorbing materials. In the center of the drum is the monochromator. Although it weighs several tons, the drum must rotate smoothly and have precision in angular setting of $\pm 0.01°$. Shielding is also needed around the analyzer crystal and the detector, but this can be much more compact. Due to the massive shielding, most modern spectrometers now move via air pads on a "tanzboden" or "dance floor." This scheme allows large masses to be moved rapidly on a cushion of air; once they are positioned, turning off the air makes the spectrometer components immovable. The separate axis units are connected together by rigid metal tubes. This style of spectrometer has a great deal of flexibility in that distances between different axes can easily be changed and the entire instrument easily positioned. Fig. 3.2 is a photograph of IN 14 at the ILL which employs these concepts.

In a reactor hall, where the detector is ~ 8–12 m from the core of the reactor, an acceptable background (neutrons reaching the detector through the shielding) is < 1 count/min, as measured with a beam striking a sample and the paths to the analyzer and detector blocked. Background can also be measured in a triple-axis mode by rotating the analyzer crystal away from

Fig. 3.2. Photograph of the IN 14 triple-axis spectrometer at the HFR reactor at Institute Laue Langevin.

the Bragg condition. This allows only stray neutrons that pass through the shielding and any incoherent scattering off the analyzer crystal to enter the counter. For instruments in a guide hall, where the detector can be 50 m from the reactor, the background is considerably less, usually < 0.1 counts/min. In an actual experiment, there is also a sample-induced background that may be from inelastic incoherent or multiple scattering.

3.2 Monochromators

Aside from the reactor, the most important component in a three-axis instrument which determines the intensity incident on the sample is the monochromator crystal. We shall now discuss the relationship governing the intensity of neutrons scattered from a monochromator. This has been treated in detail by Bacon and Lowde (1948) and Bacon (1975); here, we give a briefer discussion.

The monochromator crystal selects a specific neutron wavelength from the incident "white" beam by Bragg diffraction from a given set of lattice planes.

Bragg's Law states

$$\lambda = 2d_{hkl} \sin \theta_B. \tag{3.1}$$

In words, this states that a crystal oriented in a white beam at an angle θ_B will diffract neutrons of wavelength λ from the crystallographic planes separated by d_{hkl}, where (hkl) are the Miller indices associated with the scattering plane. For a crystal with at least orthorhombic symmetry and lattice parameters a, b, and c,

$$\frac{1}{d_{hkl}^2} = \frac{h^2}{a^2} + \frac{k^2}{b^2} + \frac{l^2}{c^2}. \tag{3.2}$$

Rewriting Bragg's Law in terms of wave-vector magnitudes, we have

$$G_{hkl} = 2k \sin \theta_B, \tag{3.3}$$

where

$$G_{hkl} = \frac{2\pi}{d_{hkl}}, \tag{3.4}$$

$$k = \frac{2\pi}{\lambda}.$$

The angular width of a Bragg-diffracted beam from a perfect single crystal is much smaller than the beam divergences with which one must work in order to get acceptable beam intensities in a neutron spectrometer. To obtain a better match, a mosaic crystal is typically used. A mosaic crystal consists of many small, perfect crystallites with small relative misorientations. If, in a given direction, the orientation of a given crystallite relative to the average is labeled Δ, then the distribution of angles is usually assumed to be approximated by a gaussian function W of width η:

$$W(\Delta) = \frac{1}{\sqrt{2\pi}\,\eta} e^{-\Delta^2/2\eta^2}. \tag{3.5}$$

[Experimentally it is more convenient to characterize the distribution by the full-width at half-maximum (FWHM), where $\eta(\text{FWHM}) = 2\sqrt{2\ln 2}\eta$ (gaussian) $\sim 2.35\eta$ (gaussian). In the following, we will use η to denote the mosaic width defined as the FWHM.] A mosaic crystal is also called an imperfect crystal, and one in which the absorption is of negligible importance is an ideally imperfect crystal. The latter case is typical for crystals chosen as neutron monochromators.

To write down formulas for the reflectivity from such a crystal, we first introduce the crystallographic quantity Q_c, given by

$$Q_c = \frac{\lambda^3 F_N^2}{v_0^2 \sin 2\theta_B}.$$
(3.6)

(Q_c should not be confused with the magnitude of the momentum transfer, $Q = |\mathbf{Q}|$.) Here, v_0 is the unit-cell volume (which is equivalent to $1/N_c$, where N_c is the number of unit cells per cm^3), and F_N is the static nuclear structure factor,

$$F_N = \sum_j b_j e^{i\mathbf{Q}\cdot\mathbf{r}_j},$$
(3.7)

where the sum is over all atoms in the unit cell, b_j is the scattering length as given in Appendix 1, and $\mathbf{r}_j$ is the position of the jth atom in the unit cell. The integrated reflectivity from a very small (volume $= \delta V$), perfect crystallite measured by rotating the crystallite in a monochromatic beam is proportional to $Q_c \delta V$.

Now consider diffraction (in reflection, as opposed to transmission) from a large, flat monochromator crystal of thickness t_0. If the crystal is ideally imperfect, with the incident and reflected angles both equal to the Bragg angle, θ_B, then the peak reflectivity is given by

$$\mathscr{R}_p = \frac{\mathscr{R}_0}{1 + \mathscr{R}_0},$$
(3.8)

where

$$\mathscr{R}_0 = \frac{Q_c t_0}{\sqrt{2\pi}\, \eta \sin \theta_B}.$$
(3.9)

For $\mathscr{R}_0 \ll 1$, $\mathscr{R}_p$ varies linearly with $\mathscr{R}_0$; however, for large $\mathscr{R}_0$, $\mathscr{R}_p$ eventually saturates at 1. The saturation is to be expected, since a crystal cannot reflect more that 100% of the incident neutrons. The depletion of the incident beam due to scattering by various crystallites is called secondary extinction; primary extinction describes the related process in a perfect single crystal due to coherent reductions in beam intensity from scattering by individual lattice planes.

Another useful quantity to calculate is $\mathscr{R}_\theta$, defined as the integrated reflectivity when the crystal is rotated in a monochromatic beam (a rocking-curve scan). It gives the angular range over which the crystal is totally reflecting. There is no simple analytic formula to describe this reflectivity; however, Bacon and Lowde (1948) found that for large values of $\mathscr{R}_\theta$, where secondary extinction is important but not saturated, $\mathscr{R}_\theta$ is given to within

5% by

$$\mathcal{R}_\theta = 0.96 \left(\frac{Q_c t_0}{\eta \sin \theta_B} \right)^{1/2}. \tag{3.10}$$

Note that in this regime the integrated reflectivity is proportional to F_N rather than F_N^2. If the monochromator is put into a white neutron beam, the integrated reflectivity as a function of wavelength, is given by

$$\mathcal{R}_\lambda = \mathcal{R}_\theta \lambda \cot \theta_B. \tag{3.11}$$

This quantity is applicable when discussing a monochromator, whereas $\mathcal{R}_p$ and $\mathcal{R}_\theta$ are more appropriate for discussions of an analyzer crystal, particularly for elastic scattering with a three-axis instrument.

From Eq. (3.10) it is clear that in order to have a large peak reflectivity, the crystallographic quantity Q_c must be large, which implies that crystals with small unit-cell volumes and large scattering lengths are desirable. Also, the best materials should have a low absorption coefficient for thermal neutrons, a large Debye temperature (i.e., a rigid lattice), and a small incoherent scattering cross section. The latter two criteria are necessary in order to keep the background low in an inelastic scattering experiment; phonon or incoherent scattering by the monochromator or analyzer yields signal at unwanted and unintended wavelengths. Some of the relevant quantities are compared in Table 3.1 for a number of materials that have been used as monochromators.

Isotopically pure ^{58}Ni would make an ideal monochromator, with a large value of F/v_0 and no incoherent scattering, if it were not so expensive to prepare. Beryllium, in principle, is the best practical material because of its large scattering cross section and small unit-cell volume; however, the small volume implies a small lattice spacing. As a result, Be is most useful for large k_i. Copper, like Be, also has large reciprocal-lattice vectors, so that these materials are frequently used for energies $E_i > 50\,\text{meV}$ ($\lambda < 1.3\,\text{Å}$).

Certain crystal lattices are advantageous for suppressing neutrons at harmonic wavelengths; an important example is the diamond structure, shared by silicon and germanium. Reflections of the form (hhh) are allowed only when h is odd or $3h = 4n$, for integer n. It follows that for the (111) reflection $F \neq 0$, while $F = 0$ for (222). Hence, when the (111) reflection is used to produce neutrons with wavelength λ, there is no contamination from neutrons with $\lambda' = \lambda/2$, because these would have to be diffracted by the (222) reflection. One must be aware that there could still be scattering by $\lambda/3$ from the (333) planes, but this occurs at 9 times the energy E_0 of the fundamental reflection, and so is likely to be relatively weak provided that $9E_0$ is in the

Table 3.1. *Some important properties of materials that are (or have been) used for neutron monochromator crystals. The last column is the ratio of the incoherent to the total scattering cross section.*

Material	Structure	Lattice parameter a (Å)	c (Å)	(hkl)	F/v_0 (10^{11} cm^{-2})	G_{hkl} (Å^{-1})	$\sigma_{inc}/\sigma_{scat}$ (%)
Beryllium	hcp	2.2854	3.5807	(002)	0.962	3.5095	0.02
				(110)	0.962	5.4985	
Iron	bcc	2.86645		(110)	0.802	3.1000	3.4
Zinc	hcp	2.6589	4.9349	(002)	0.376	2.5464	1.9
PG^a	layer	2.4612	6.7079	(002)	0.734	1.8734	0.02
				(004)	0.734	3.7467	
Niobium	bcc	3.3008		(200)	0.392	3.8071	0.04
Nickel (^{58}Ni)	fcc	3.52394		(220)	1.316	5.0431	0
Copper	fcc	3.61509		(220)	0.653	4.9159	6.8
Aluminum	fcc	4.04964		(220)	0.208	4.3884	0.55
Lead	fcc	4.9505		(220)	0.310	3.5898	0.03
Silicon	diamond	5.43072		(111)	0.147	2.0039	0.2
				(220)	0.207	3.2724	
				(311)	0.147	3.8372	
Germanium	diamond	5.65776		(111)	0.256	1.9235	2.1
				(220)	0.362	3.1411	
				(311)	0.256	3.6832	

^{a} PG = pyrolytic graphite.

high-energy tail of the spectrum of thermally-moderated incident neutrons. For reasons that will become clear after we discuss filters, the problem of reducing second-order contamination from a monochromator that diffracts $\lambda/2$ neutrons can place severe restrictions on the choice of incident neutron energy. Therefore, use of a monochromator reflection that is inherently free of unwanted harmonics provides considerable freedom in the choice of incident energy, which can allow for fine tuning of the energy and momentum resolution in a simple manner.

The throughput of a spectrometer depends not only on the peak reflectivity of the monochromator, but also on the angular width of the reflection, which corresponds to the mosaic width, η. The major limitation encountered in using many of the materials listed in Table 3.1 is that their as-grown crystalline forms are usually too perfect, i.e., η is too small (typically $\eta < 1'$); as a result, the integrated reflectivity is very low. The challenge over the last several decades has been to change the mosaic characteristics in these crystals in a controlled manner and with sufficient homogeneity so that the Bragg reflection maintains a simple gaussian shape. This has been a difficult task, but considerable success has been achieved under the guidance of A. Freund, formerly at the ILL (Freund, 1976). The method used for controlling the mosaic is to introduce dislocations in a crystal by plastically deforming it at temperatures near the melting temperature. Unfortunately, the dislocations concentrate in low-angle grain boundaries, which gives rise to non-gaussian mosaic distributions in large crystals, as well as spatial variations in the mosaic across the crystal. The situation in Be is even worse, since it is not possible, at present, to grow large crystals reproducibly. There are a handful of "legendary" Be crystals in use at existing facilities, but attempts to reproduce their properties have met with limited success.

A new process to introduce a spatially homogeneous, but anisotropic, mosaic structure has been developed at Brookhaven National Laboratory (Vogt *et al.* 1994). In this technique, single-crystal semiconductor wafers, typically 0.3 mm in thickness, are heated and then repeatedly deformed and flattened. They are then reassembled to form a composite of any desired thickness by using tin to bond the wafers together. The technique has been successfully applied to Si and Ge crystals, where enhancements of the reflectivities by factors greater than 2 have been achieved. Of equal importance, the mosaic distribution is uniform. Anisotropy of the mosaic width is beneficial for matching the differing beam divergence requirements in the horizontal and vertical directions. Twenty-four germanium composite crystals were used in a focusing monochromator built for the high-resolution powder diffractometer at Brookhaven.

Even with the above technique, the integrated reflectivity is still too low for routine use as a monochromator in a three-axis inelastic instrument, since the wavelength acceptance ($\Delta\lambda/\lambda$) is too small. An alternative way of increasing $\Delta\lambda/\lambda$ without sacrificing peak reflectivities is to tilt individual wafers with respect to each other. Such "fanned" arrays with a controlled misalignment have yielded peak reflectivities approaching 80% that of pyrolytic graphite monochromators (see below).

Another difficulty in using single crystals as monochromators results from double Bragg scattering (for more discussion, see Chap. 7). This occurs when two crystallographic planes simultaneously satisfy the condition for Bragg scattering as given in Eq. (3.1). The probability of this occurring increases with the size of the unit cell and the illuminated volume. It not only gives an increase in background, but it also causes large variations in the measured R^λ as a function of λ because the conditions for multiple scattering vary

Table 3.2. *The performance at* $\lambda = 1.27$ Å *of different monochromators (Riste and Otnes, 1969). PG stands for pyrolytic graphite.*

Crystal	Reflection	η (')	$\mathscr{R}_\theta$ (')	$\mathscr{R}_\theta/\eta$	$\mathscr{R}_\lambda$ (0.01 Å)	$\mathscr{R}_p$
Be	002	22	11	0.5	1.1	0.42
Cu	111	22	4.7	0.19	0.53	0.14
Zn	002	34	13.6	0.39	1.9	0.31
Ge	111	18	4.8	0.27	0.9	0.22
PG	002	68	58	0.86	8.7	0.74

as λ varies. In addition, the reflectivity is sensitive to rotation of the crystal about the scattering vector, since this will cause reflections competing with the desired one to move into or out of the scattering plane.

A nearly ideal material for use as a monochromator is pyrolytic (or oriented) graphite (PG) since many of the problems enumerated above are avoided with the use of this material (Riste and Otnes, 1969). Graphite has a hexagonal, layered structure. Pyrolitic graphite (PG) has highly preferred orientation of the (00*l*) planes, but all other (*hkl*) planes are aligned at random, giving rise to powder peaks in a diffraction pattern, similar to those of a polycrystalline specimen. The relevant characterization for PG is the mosaic width for (00*l*) reflections; "good" monochromators have $\eta \lesssim 0.5°$. Table 3.2 gives a comparison of the PG (002) reflection with other typical monochromator crystals tested by Riste and Otnes (1969). They used a relatively poor PG crystal with a mosaic width of $1.13°$ and thickness $t_0 = 3.7$ mm. It can be seen that it outperforms the other materials by better than a factor of 2 in every measure.

Figure 3.3 shows the measured peak reflectivity, $\mathscr{R}_p$, over the energy range of 4.5–40 meV (4.3 Å $> \lambda > 1.4$ Å) for the PG (002) reflection of a crystal with $t_0 = 1.6$ mm and $\eta = 0.56°$ $(0.63°)$ at 25 meV (4 meV) (Shapiro and Chesser, 1972); the absorption was less than 2%. The peak reflectivity is about 80% at low energies, and decreases to about 60% at 40 meV. There are pronounced dips at 15.3 and 23.6 meV which are due to double Bragg scattering. Since the planes perpendicular to the *c*-axis are randomly oriented with respect to each other, the Bragg points of these planes can be considered to be Bragg rings about the (00*l*) axis. The scattering plane will intersect all of these rings as shown in Fig. 3.4. For certain values of the energy, the Ewald sphere (see §1.3.1) will intersect one or more rings; this is the condition for double Bragg scattering as explained in more detail in Chap. 7. For these

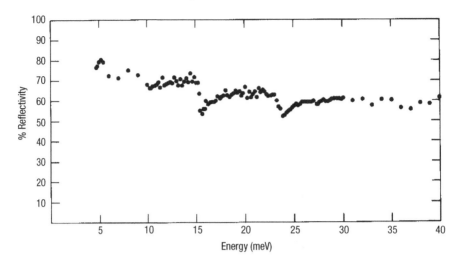

Fig. 3.3. Measured peak reflectivity, $\mathscr{R}_p$, as a function of energy (from Shapiro and Chesser, 1972).

energies, part of the beam will be scattered at an angle other than that of the primary scattering. This part of the beam will most likely not enter the detector and therefore a spuriously low counting rate will be obtained. The two largest dips correspond to reflections from (112) and (114) for energies at 15.3 meV ($k_i = 2.72\,\text{Å}^{-1}$) and 23.6 meV ($k_i = 3.32\,\text{Å}^{-1}$), respectively.

The Ewald sphere will first touch the Bragg rings within the scattering plane. As the energy increases, the Ewald sphere becomes larger and it will pass through the rings. Since the sphere must pass through the points (000) and (002), and since the rings are concentric about the (00l) axis, a ring can never be wholly contained within the sphere. Hence, once the conditions are met for the double scattering process, it will continue to "eat away" part of the beam. The dips, therefore, are expected to be sharp on the low-energy side (where the Bragg ring is tangent to the Ewald sphere) and tail off on the high-energy side as the sphere moves through the rings. As the energy increases, the superposition of many of these "tails", coupled with the decrease of λ, leads to an overall decrease in the reflectivity. Below 5 meV double scattering is of little significance, and the reflectivity behaves like that of an ideal mosaic crystal with secondary extinction.

3.2.1 Focusing monochromators and analyzers

One method of increasing the flux on a sample is to employ focusing of the beam (Meier-Leibnitz, 1967; Rustichelli, 1969). For neutrons, the source is

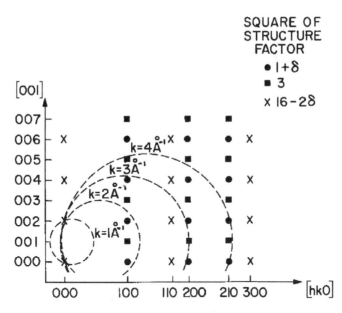

Fig. 3.4. Scattering plane for the (00*l*) reflection of pyrolytic graphite. The points shown (except those along [00*l*]) actually represent the intersection of the plane with Bragg rings about the [00*l*] axis. Dashed circles are the intersections of the Ewald sphere for (002) at various wave vector magnitudes k. The parameter δ in the structure factor is essentially zero (from Shapiro and Chesser, 1972).

usually much larger in spatial extent than an X-ray source. The focusing for X-rays usually involves a monochromator bent to a cylindrical surface whose axis is perpendicular to the plane of scattering. This horizontal focusing was initially suggested for neutron scattering by several persons in the 1960s, and was later tested by Scherm *et al.* (1977). Since good Q resolution within the scattering plane is frequently desired, while poor resolution can be tolerated perpendicular to the plane, vertical focusing is more commonly used. Riste (1970) suggested a vertical focusing of neutrons, where the cylindrical, focused surface has its axis in the scattering plane. Because of the requisite shielding of the reactor and the monochromator, the sample axis is typically 5 m or more from the core. The natural vertical collimation of the beam incident on the sample is, therefore, generally better than actually needed. Vertically-focusing monochromators will increase the vertical divergence and allow for an enhancement in flux (typically $\sim 4\times$) without seriously affecting the in-plane momentum and energy resolution of the instrument.

   A vertically bent crystal can also be used as an analyzer. The same result can be achieved by using a larger detector or a smaller sample-to-detector

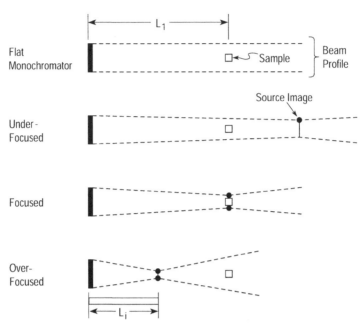

Fig. 3.5. Beam profile as a function of degree of focusing. Diagrams represent from top to bottom $R = \infty$ (flat monochromator), R greater than focusing value, R satisfying focusing condition, and R less than focusing value (from Nunes and Shirane, 1971).

distance; however, a larger detector means a larger background. With a bent analyzer one can use a smaller detector, resulting in a lower background.

A vertically bent monochromator can be treated as a cylindrical mirror and the focusing properties are well described by simple geometric optics. If the source is at a distance L_0 (see Fig. 3.1) from a concave bent monochromator with radius of curvature, R, and the monochromator reflects a beam at an angle θ_M, then an image will be formed at a distance L_i. The relationship between these quantities is

$$\frac{1}{L_0} + \frac{1}{L_i} = \frac{2 \sin \theta_M}{R}. \tag{3.12}$$

The focusing condition is achieved when $L_i = L_1$, where L_1 is the monochromator to sample distance (see Fig. 3.5). Since L_0 and L_1 are usually fixed, the focusing condition is energy-dependent because of the dependence on $\sin \theta_M$. Thus as the energy is increased, the focusing distance, L_i, will vary and the beam will range from being under-focused, to focused, to over-focused as shown in Fig. 3.5.

The features of practical interest are the effective gain in flux one will achieve using a focused monochromator and its effect on the instrumental resolution. Nunes and Shirane (1971) have provided an analysis of this problem, and compared the calculations with measurements. To start, one has to estimate the vertical spread of the neutron beam at the sample. If the height of the source is h_s, then the height of the image, h_i, will be

$$h_i = h_s \left(\frac{L_i}{L_0} \right). \tag{3.13}$$

If $L_i = L_1$ the system is focused, and the sample experiences the maximum flux. Typically, $L_1 < L_0$, and h_i is less than h_s. If the sample height is less than or equal to the image height, the flux increase due to focusing is translated directly into an intensity increase. If $L_i \neq L_0$, the height of the beam at the sample can be calculated. If h_M is the height of the curved monochromator, the height of the beam at the sample, H_{curv}, will be

$$H_{curv} = \begin{cases} h_M - (h_M - h_i) \left(\dfrac{L_1}{L_i} \right), & L_i \geq L_1; \\[2ex] h_i + (h_M - h_i) \left(\dfrac{L_1 - L_i}{L_i} \right), & L_i \leq L_1. \end{cases} \tag{3.14}$$

For a flat monochromator, $R = L_i = \infty$, and the height at the sample, H_{flat}, is

$$H_{flat} = h_M + h_s \left(\frac{L_1}{L_0} \right). \tag{3.15}$$

If the neutron flux is uniform and inversely proportional to the beam height, the gain in neutron flux at the sample from use of a focusing monochromator is given by the flux ratio,

$$P = \frac{H_{flat}}{H_{curv}}. \tag{3.16}$$

The maximum increase in flux occurs for perfect focusing,

$$P_{max} = \frac{h_M + h_s \left(\dfrac{L_1}{L_0} \right)}{h_s \left(\dfrac{L_1}{L_0} \right)}. \tag{3.17}$$

For example, if $h_M \approx h_s$ and $L_1/L_0 = \frac{1}{2}$, the gain in flux is ~ 3 with respect to a flat monochromator; even greater gains can be achieved with $h_M > h_s$. The radius of curvature required for focusing can be found by using Eq. (3.12) with $L_i = L_1$, calculating $\sin \theta_M$ from the neutron wavelength, and solving for R.

Figure 3.6 shows a comparison of the calculated flux ratio, using Eq. (3.16), and the observed incoherent scattering from a vanadium cylinder. The

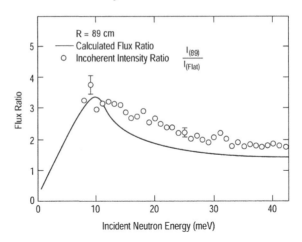

Fig. 3.6. Calculated flux ratio [Eq. (3.16)] compared with observed incoherent scattering using an $R = 89$ cm monochromator. Under-focusing occurs at the high-energy side of the peak and over-focusing at the lower-energy side (from Nunes and Shirane, 1971).

monochromator was a 5-cm-high piece of bent pyrolytic graphite (PG) with fixed radius of curvature $R = 89$ cm. Measurements were performed on the H7 beam line at Brookhaven National Laboratory's High Flux Beam Reactor (HFBR). For the dimensions of this spectrometer, perfect focusing occurs at approximately 11 meV, with a maximum flux ratio of 3.3. Assuming the height of the source to be 8.9 cm, the image height at focusing will be 2.2 cm. Any sample less than 2.2 cm in height should realize an intensity improvement of a factor of 3.3 with respect to that of a normal flat monochromator. The calculated flux ratio falls off rapidly for energies less than the focusing energy. This is the case where over-focusing ($L_i < L_1$) occurs. For larger energies, under-focusing occurs and the curve falls off less rapidly. It is quite clear that the latter case is the preferred one. [Currat (1973), using a somewhat more sophisticated analysis, has shown that much better agreement with the observed intensity ratio can be obtained.]

The original vertically-focusing PG monochromators typically involved a single piece of PG, approximately 10×7 cm^2 in size, that was hot-pressed to a shape with a fixed radius of curvature. These are prohibitively expensive to manufacture, so it is now customary to construct a "venetian blind" type of monochromator, where small strips of PG (or other monochromator crystals) are mounted on a flexible backing which can be curved by an electromechanical device. Another variation involves mounting each crystal on a plate with its own adjusting screws to ensure a proper alignment. This "venetian blind" mount allows one to vary the radius of curvature to account for changes in energy and maintain optimum focusing. It also allows one to build larger monochromators and focus more of the neutrons leaving the reactor onto a sample.

In addition to the vertical focusing of the neutrons, horizontal focusing is now frequently employed in the analyzer as well as in the monochromator (Scherm *et al.*, 1977; Bührer *et al.*, 1981). Whereas vertical focusing involves only the divergence perpendicular to the scattering plane and has a minimal effect on the energy resolution, horizontal focusing with a crystal monochromator influences simultaneously the imaging properties and the wavelength resolution in the scattering plane by introducing a correlation between the direction and the wavelength of the diffracted beam. This changes the resolution function in a complicated manner, but the advantage is that the instrumental resolution can be tailored according to specific requirements by varying parameters like the curvature of the reflecting plane, using asymmetric cut crystals or a crystal with a lattice gradient.

A recent development is the employment of simultaneous vertically and horizontally-focusing (double-focusing) monochromators in combination with horizontally-focusing analyzers. An experiment measuring phonons in copper crystals has shown intensity gains of up to a factor of 100 over measurements using a flat monochromator and analyzer, with less than a factor of 2 change in energy widths (Bührer, 1994). A more complete discussion of focusing monochromators can be found in the *Proceedings of the Workshop on Focusing Bragg Optics* (Magerl and Wagner, 1994).

3.2.2 Resolution effects

A curved, vertically-focusing monochromator increases the vertical divergence of the beam incident upon the sample, which affects the triple-axis resolution function. The net effect, however, is not necessarily large. The resolution function is determined jointly by the monochromator and analyzer systems, and the vertical divergence of the analyzer section is usually much greater than the monochromator section. In fact, the increase in vertical divergence obtained with a curved, vertically-focusing monochromator provides a better matching between the monochromator- and analyzer-system parameters, which produces a more efficient instrument without seriously degrading the resolution.

We will discuss the resolution function in some detail in Chap. 4. For the time being, we simply note that, to first order, the width ΔQ_z of the distribution for momentum-transfer components in the vertical direction is decoupled from the resolution widths for the in-plane components (ΔQ_x, ΔQ_y) and for the energy transfer, $\Delta\hbar\omega$. For many purposes, the effect of a curved, vertically-focusing monochromator on the calculated resolution function can be taken into account by using an appropriately increased value for the monochromator's vertical mosaic width. The only observable impact is on ΔQ_z, with essentially none on ΔQ_x, ΔQ_y, or $\Delta\hbar\omega$. A vertically-focusing analyzer can be accommodated in a similar fashion, although the larger effect on the vertical resolution may not be described as accurately. In some instances, it may be desirable to take more careful account of spatial correlations that occur with focusing optics, and this can be done using the approach of Popovici, Stocia, and Ioniță (1987).

The effects on the resolution due to vertical focusing have been tested using nearly perfect single crystals ($\eta < 0.01°$) of BaTiO$_3$ and silicon; for BaTiO$_3$, the results are shown in Fig. 3.7 (Nunes and Shirane, 1971). The upper part shows energy scans through a Bragg peak of BaTiO$_3$ for a bent monochromator and for a flat monochromator. Although the intensities differ, the energy widths are nearly identical. The bottom part demonstrates the dependence of the Bragg intensity on the tilting angle of the sample. From this, ΔQ_z is calculated as the sample scattering-vector magnitude, Q, times the FWHM of the tipping curve expressed in radians. Again, the widths are nearly identical.

Figure 3.7 demonstrates another important consideration. The flux gains calculated and confirmed in Fig. 3.6 were for incoherent scattering which is distributed isotropically in Q space. Similar gains are achieved when measuring phonon intensities from a crystal and diffraction peaks in a powder. However, these gains are not realized with Bragg scattering from single crystals, as shown in Fig. 3.7 which depicts a flux ratio of only 1.5, compared to 3.3 for incoherent scattering. This is due to the more stringent geometrical requirements of the Bragg-scattering process. The vertical divergence of the incoherently scattered beam is independent from that of the incident beam, so incoherent intensity depends only upon the absolute number of neutrons striking the sample and not their incident direction. To a lesser extent, this is also true of phonon scattering. The divergence of the Bragg-scattered beam is, however, directly determined by that of the incident beam, and as the detected Bragg intensity depends upon both beam intensity and divergence the intensity ratio expression for Bragg scattering is more complicated than Eq. (3.16). Generally, the highest Bragg intensity gain will be observed for an under-focused incident beam for which the vertical divergence has not increased sufficiently to reduce the effect of the increased flux at the sample.

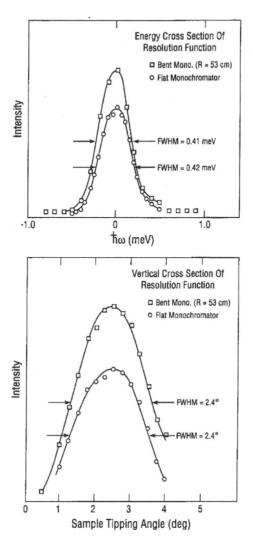

Fig. 3.7. Scans performed using the (004) reflection of a perfect BaTiO$_3$ crystal with an incident energy of $E_0 = 38$ meV and horizontal collimations of 20′-20′-20′-40′ (from before the monochromator to after the analyzer). (*Top*): constant **Q** scans. (*Bottom*): tilting curves from which ΔQ_z is calculated (from Nunes and Shirane, 1971).

3.2.3 *Double monochromators*

In a standard triple-axis spectrometer with a single monochromator crystal (see Fig. 3.1), the entire spectrometer must move when the incident energy is changed. If a large range of incident energies is to be available, then the spectrometer requires a substantial floor area for motion. Such a requirement

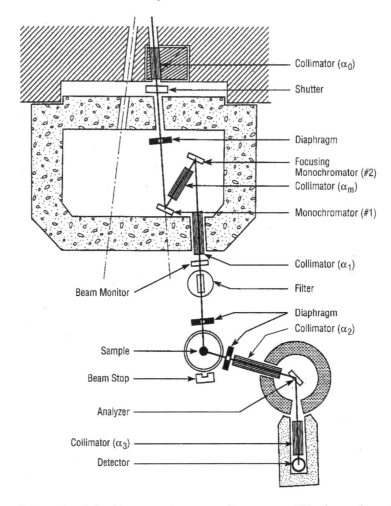

Fig. 3.8. Schematic of double-monochromator instrument 4F2, situated on the cold source at the Orphée reactor, Laboratoire Léon Brillouin, Saclay, France.

is sometimes in conflict with the needs of neighboring instruments. One solution is to create a monochomatized beam with a fixed direction independent of energy using a two-crystal (double) monochromator. The disadvantage of a double-monochromator instrument is the reduction of intensity resulting from the additional reflection and the increased neutron flight path.

Figure 3.8 shows a schematic of the 4F2 instrument located on the cold source at Laboratoire Léon Brillouin in France. Inside the shielding are two monochromators with a collimator between them. In this design, the first monochromator (#1) is flat and the second one (#2) is curved and vertically focused on the sample. The wavelength is changed by rotating the monochromator assembly about the first crystal and translating crystal #2 along the final beam direction so that the beam always leaves the shielding through the fixed exit hole.

For an instrument with a single monochromator, the configuration of horizontal collimators is typically specified as α_0-α_1-α_2-α_3, where α_j are replaced by the respective angular divergences in minutes of arc. The angular divergences are used in calculating the instrumental resolution function (see Chap. 4). It is desirable to be able to apply the same formulas for the resolution function in the case of a double monochromator. Pynn and Passell (1974) have shown that one can do so by adopting the following definitions:

(i) effective horizontal exit collimation α_1^{eff}:

$$\frac{1}{(\alpha_1^{\text{eff}})^2} = \frac{1}{\alpha_0^2} + \frac{1}{\alpha_1^2} \tag{3.18}$$

(ii) effective horizontal in-pile collimation α_0^{eff}:

$$\alpha_0^{\text{eff}} = \alpha_m \tag{3.19}$$

(iii) effective horizontal mosaic width of the monochromator η_H^{eff}:

$$\frac{1}{(\eta_H^{\text{eff}})^2} = \frac{1}{\eta_{H1}^2} + \frac{1}{\eta_{H2}^2} \tag{3.20}$$

(iv) effective vertical mosaic width of monochromator η_V^{eff}:

$$\frac{1}{(\eta_V^{\text{eff}})^2} = \frac{1}{\eta_{V1}^2} + \frac{1}{\eta_{V2}^2} \tag{3.21}$$

The notation of α_j is given in Fig. 3.8, and the additional subscripts 1 or 2 attached to η_H and η_V denote the first or second monochromator, respectively.

3.3 Collimators

Unlike the highly-collimated beam of X-rays produced by a synchrotron, the moderated neutrons from a reactor emerge in all directions. While Bragg diffraction by the monochromator and analyzer crystals serve to provide some constraint on the angular divergence of the beam, it is generally desirable to have additional adjustable control. Control of the beam divergence within the horizontal scattering plane is typically achieved through the use of Soller collimators, devices first employed in X-ray work (Soller, 1924). A Soller collimator consists of parallel absorbing plates of length L and height h separated by a distance a. The ideal transmission function is triangular, with a FWHM of $\alpha = a/L$ in the horizontal plane, and $\beta = h/L$ in the vertical. For resolution calculations, where multiple collimators and crystals are involved, the central limit theorem implies that a gaussian approximation to the triangular transmission function should be reasonable, with gaussian widths $\alpha' = \alpha/\sqrt{8 \ln 2}$ and $\beta' = \beta/\sqrt{8 \ln 2}$ (Sailor *et al.*, 1956). Typical values for the horizontal divergence are in the range $0.1° \lesssim \alpha \lesssim 1.5°$.

The efficiency of a Soller collimator depends upon a number of factors: (i) the uniformity of the spacing of the blades throughout the collimator, (ii) the neutron absorption cross section of the blade or septa material, (iii) the thickness of the blades, and (iv) the straightness of the blade edges at the entrance and exit to the collimator. For high transmission, points (i), (iii), and (iv) must be optimized, and (ii) is important for minimizing the background.

Early collimators consisted of thin steel blades coated with the neutron absorber cadmium. A rectangular frame had carefully machined slots into which the blades could easily slide. The width and length of the open slots defined by the blades determined the angular divergence of the beam in the horizontal direction. By removing or adding blades, the angular divergence could easily be changed. Later, a high-efficiency Soller collimator was developed (Carlile, Hey, and Mack, 1977) using a thin (0.025 mm) plastic (Melinx) film stretched over a stainless steel or aluminum frame. The thickness of the frame determines the spacing between the septa. Two side plates are used to support the pack of blades. The plastic films are coated with a neutron-absorbing paint containing gadolinium oxide, which has a high neutron capture cross section. These collimators satisfy all the conditions listed in the paragraph above and have essentially a triangular transmission function with a peak transmission exceeding 95%. Since the blades are so thin, this type of collimator is desirable for a fine collimation; e.g., $\alpha < 0.3°$. (In a conventional collimator with a convenient length, the transmission suffers as the spacing of the blades becomes comparable to their thickness.) The film collimators are ideal since, even for fine collimation, the blades fill up a small fraction of the collimator volume. One disadvantage of this type of collimator is that the spacing between the blades cannot be changed, so that an entirely different collimator is needed for each choice of angular divergence.

The angular beam divergences in the various arms of the spectrometer are significant input parameters for calculating the resolution function (see Chap. 4). It is important to recognize that these divergences are not always limited by the collimators. When a small sample is measured with coarse collimation (which is especially common in the vertical direction), the beam divergence may be limited by the sample dimension and the size of the monochromator (or analyzer) or an intervening, beam-defining mask. For example, suppose that a sample of height c is a distance S from a beam-defining mask of height d. The effective vertical divergence is then $\tilde{\beta} = (c + d)/2S$. When this value is compared with β for the collimator along the same path, the smaller of the two should be used in the resolution calculation.

3.4 Filters

As we have noted above, the monochromator crystal selects a very narrow wavelength band from a broad incident spectrum using Bragg diffraction. Following Bragg's Law, Eq. (3.1), a particular angular setting of the monochromator determines the ratio $\lambda/2d_{hkl}$ for a reflection (hkl). It follows that if the reflection (nh, nk, nl) is also allowed, where $d_{nh,nk,nl} = d_{hkl}/n$, then for the same setting the monochromator will also diffract neutrons at wavelength λ/n. The presence of such higher-order neutrons results in extra peaks in a diffraction pattern or an inelastic spectrum. These features can play havoc in interpreting measurements, especially in problems involving phase transitions where superlattice peaks are expected. Thus, filters are needed to effectively eliminate the unavoidable higher-order neutrons and to reduce the background.

The total cross section determining the attenuation of neutrons traversing a crystalline solid is given by the sum:

$$\sigma_T = \sigma_{abs} + \sigma_{el} + \sigma_{inel}. \tag{3.22}$$

The first term, σ_{abs}, is the absorption due to nuclear capture processes and varies as $E^{-1/2}$ (the inverse of the neutron velocity), with sharp spikes in the absorption due to various resonant nuclear absorption processes (which can be useful for filtering, see below). σ_{el} corresponds to the elastic Bragg scattering in the solid. The maximum wavelength λ_{cutoff} which can be diffracted by a crystalline solid (the Bragg cutoff) is equal to twice the largest d-spacing. For $\lambda > \lambda_{cutoff}$, σ_{el} is zero. For polycrystalline samples, there is a steep increase in σ_{el} immediately above the cutoff energy, followed by wiggles which are due to successively activated reflections. Below, we shall discuss the use of this cutoff energy for low-energy filters. At larger energies, σ_{el} gradually decreases, approaching zero for high energies. σ_{inel} is the inelastic scattering due to single-phonon and multiphonon processes in the solid; it is energy- and temperature-dependent. The inelastic cross section increases with energy, with multiphonon scattering dominating at high incident energies.

Freund (1983) has calculated the transmission properties of several materials and gives a semi-empirical form with which one can calculate the total transmission as a function of neutron energy E. Excluding the nuclear resonant effects, and neglecting Bragg scattering, he finds

$$\sigma_T = C_1 E^{-1/2} + C_2(T) E^{-1/2} + C_3(T)\{1 - \exp[-C_4(T)E]\}. \tag{3.23}$$

(The coefficients are obviously material dependent.) The first term on the right-hand side corresponds to nuclear absorption, and the other two terms correspond to single-phonon and multiphonon processes, respectively.

For a material to be an effective filter, σ_T must be small in the energy range of interest, and large at the higher energies which need to be filtered out of the beam. We shall now give examples of filters designed to use either the attenuation processes expressed in Eq. (3.22), resonant absorption, or Bragg-scattering effects.

3.4.1 Fast neutron filters

Three-axis spectrometers are typically used in the thermal energy regime ($E_i < 100\,\mathrm{meV}$). Neutrons of higher energies are always present and are very difficult to shield against. They are one of the main contributors to the background in any experiment. The newer reactors have tangential beam tubes and heavy-water moderators which reduce the number of these unwanted higher-energy, or fast, neutrons; however, in some older reactors, and even the newer reactors with beam tubes that look more directly at the core, there are a large number of fast neutrons that need to be eliminated.

Filters for fast neutrons make use of σ_inel to strongly attenuate the flux at high energies. Single crystals are chosen in order to minimize the elastic cross section. Various materials have been used, including Bi, Si, quartz (SiO_2) and sapphire (Al_2O_3); figure 3.9 shows the total cross section measured for sapphire as a function of energy (Freund, 1983), with fitted curves based on Eq. (3.23). It can be seen that the cross section is small in the 20–50 meV region, where nuclear absorption and single-phonon processes dominate, and it increases by nearly a factor of 10 for $E > 100\,\mathrm{meV}$. The large variation with energy in the 20–60 meV regime is due to Bragg scattering from various crystallographic planes. This can be reduced by tuning the crystal or by using a crystal with a higher degree of perfection. Tennant (1988) measured the transmission of a super-optical quality single crystal of sapphire as a function of energy and temperature. His results are shown in Fig. 3.10 as a function of wavelength. The transmission is high and varies slowly over the thermal- and cold-neutron regime ($\lambda > 1.0\,\text{Å}$). In the epithermal regime,

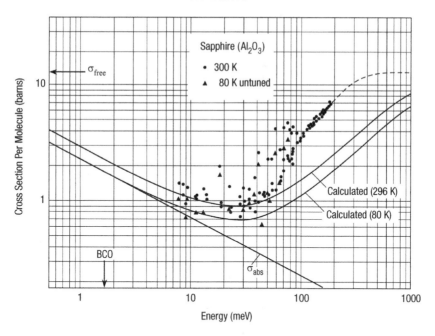

Fig. 3.9. Total cross section of sapphire. The data are from Nieman, Tennant, and Dolling (1980); the curves are fits to Eq. (3.23) by Freund (1983).

below $\lambda = 0.4\,\text{Å}$ ($E > 500\,\text{meV}$), the transmission is less than 3%. Cooling the crystal has only a small effect on the transmission. This material is useful as a filter for generally reducing the flux of fast neutrons, and is also an effective filter for $\lambda/3$ when λ is less than $1.2\,\text{Å}$ ($E_0 > 55\,\text{meV}$). Sapphire is superior to quartz and silicon as a filter for fast neutrons (Nieman, Tennant, and Dolling, 1980). Recent measurements by Mildner and Lamaze (1998) show that the transmission characteristics of single-crystal sapphire are not changed by several years of neutron irradiation.

3.4.2 Resonance filters

In the higher-energy regime ($E > 50\,\text{meV}$), filters based upon nuclear resonances can be used. All nuclei have nuclear resonances which are determined by their nuclear structure, and measurements of these have been tabulated by Mughabghab, Divadeenam, and Holden (1981) [this is the latest version of the original "Barn Book" of Hughes and Harvey (1955)]. A number of elements having strong resonances in the epithermal regime can be utilized as filters of higher-order neutrons. Table 3.3 gives a list of nuclei which have resonances in the wavelength range 0.24–0.53 Å (energy range 1.42 eV–0.29 eV)

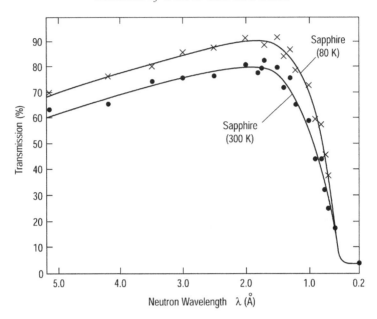

Fig. 3.10. Transmission of 10.2-cm-thick sapphire filter measured at 300 K and 80 K as a function of wavelength (from Tennant, 1988).

which make suitable $\lambda/2$ filters for λ in the range 0.48–1.06 Å (363–72 meV), or $\lambda/3$ filters for λ in the range 0.72–1.59 Å (161–32 meV). In the table we list for each element the total cross section at the resonance ($\lambda/2$) and at the wavelength of twice the resonance (λ), the ratio of which gives a figure of merit for these materials for use as filters in the high-energy regime. There are no nuclei with strong resonances at wavelengths longer than 0.53 Å so such higher-order contamination must be eliminated by filters relying on elastic Bragg scattering.

3.4.3 *Bragg-scattering filters*

As mentioned above, the maximum wavelength for which Bragg scattering can occur is

$$\lambda_{\text{cutoff}} = 2d_{\text{max}}, \tag{3.24}$$

where d_{max} is the largest d-spacing (smallest Miller indices) of the planes in the crystal for which the structure factor is finite. For wavelengths greater than λ_{cutoff} the material is transparent. This typically occurs in the 4–8-Å range, so filters based on elastic scattering are useful only in this subthermal regime. Figure 3.11 shows the cross section in the vicinity of λ_{cutoff} for a

Table 3.3. *The total cross section σ_T and the resonance energy of a number of isotopes in common use as neutron filters (Freund and Forsyth, 1979).*

Element	Resonance (eV)	$\sigma_T(\lambda/2\text{-resonance})$ (barns)	λ (Å)	$E(\lambda)$ meV	$\sigma_T(\lambda)$ (barns)	$\dfrac{\sigma_T(\lambda/2)}{\sigma_T(\lambda)}$
^{239}Pu	0.29	3800	1.06	72.5	500	7.6
^{231}Pa	0.39	4900	0.92	97.5	116	42.2
Eu	0.46	10100	0.84	115	1050	9.6
Er	0.46	2300	0.84	115	125	18.4
	0.58	1500	0.75	145	127	11.8
^{229}Th	0.61	6200	0.73	152.5	< 100	> 62.0
Ir	0.66	4950	0.70	165	183	27.0
^{240}Pu	1.08	115000	0.55	270	145	79.3
Hf	1.10	5000	0.55	275	58	86.2
Rh	1.27	4500	0.51	317.5	76	59.2
In	1.45	30000	0.48	362.5	94	319

number of materials. Polycrystalline materials are used for this type of filter since the orientation of planes is not a concern for $\lambda > \lambda_{\text{cutoff}}$. For $\lambda < \lambda_{\text{cutoff}}$, one desires as many orientations of the atomic planes as possible to scatter the unwanted neutrons. Of the materials shown in Fig. 3.11, the two most useful as low-energy band-pass filters are Be and BeO, with energy cutoffs of 5.2 and 3.7 meV (4.0 and 4.7 Å), respectively.

Figure 3.12 shows the transmission as a function of wavelength for a 15-cm length of polycrystalline Be placed in the incident beam at the NRU reactor at Chalk River (Tennant, 1988). The transmission measured at 1.5 Å (36 meV) is 2.6×10^{-5} (not shown), whereas that at 4.0 Å is 71% at $T = 80$ K. Figure 3.12 also demonstrates that cooling the polycrystalline filter greatly increases the transmission for $\lambda > \lambda_{\text{cutoff}}$ but does not affect the cutoff wavelength. Cooling reduces the amount of inelastic scattering by phonons, and this increases the transmission without appreciably changing d_{max}.

When working with a fixed incident energy of 5 meV or less, it is standard practice to place a cooled Be filter in the incident beam. An alternative application is to use a Be filter to energy-analyze the scattered beam, with no crystal analyzer. In this mode, the incident energy is varied, and all scattered neutrons with $E < E_{\text{cutoff}}$ are detected. The large band pass and solid angle for collecting scattered neutrons provide a considerable efficiency. The disadvantage is that Q is not constant over the energy-transfer range. Since $k_f \sim 1\,\text{Å}^{-1}$, the magnitude of Q is nearly equal to k_i, which varies with the average energy transfer. This method has been used successfully in

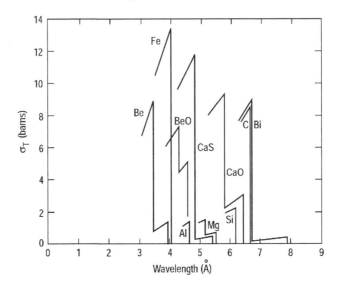

Fig. 3.11. Total neutron cross section as a function of wavelength in the vicinity of the Bragg cutoff wavelength for a number of polycrystalline materials (from Freund and Forsyth, 1979).

molecular spectroscopy for hydrogen-containing systems where incoherent scattering is measured and density of states information is desired (Richter and Shapiro, 1980). A recent application has been to measure the density of states in C_{60} (Copley, Neumann, and Kamitakahara, 1995).

Pyrolytic graphite (PG) is by far the most useful filter of higher-order neutrons in the thermal energy regime. A PG filter is placed in the beam with its c-axis parallel to the beam. Bragg scattering can take place only for k values which satisfy the relation

$$2k \sin(90° - \phi_{hkl}) = G_{hkl}, \tag{3.25}$$

where ϕ_{hkl} is the angle between the reciprocal-lattice vectors $\mathbf{G}_{hkl}$ and the c-axis. Those incident neutrons with wave vectors which satisfy this relation are scattered out of the incident beam. Figure 3.13 shows the total cross section (Loopstra, 1966) for neutrons traveling parallel to the c-axis for a 1-cm-thick piece of PG. The maximum cross section occurs around $E = 60\,\text{meV}$, while at $E/4$ ($= 15\,\text{meV}$) the cross section approaches a minimum. It follows that this crystal is a good filter transmitting 15 meV neutrons and attenuating neutrons at $\lambda/2$. Figure 3.14 presents similar results for a 5-cm-thick PG filter in the form of transmission for λ (E), $\lambda/2$ $(4E)$, and $\lambda/3$ $(9E)$ over the range $10 < E < 45\,\text{meV}$. Ideally, one would like to have a large transmission for λ together with a small transmission at $\lambda/2$ and $\lambda/3$. This is satisfied in the

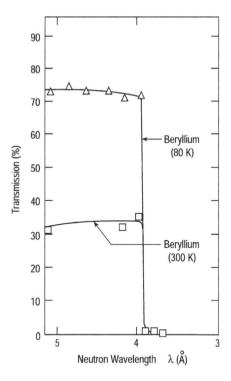

Fig. 3.12. Transmission of a 15-cm length of polycrystalline beryllium as a function of wavelength measured at $T = 300$ K and 80 K (from Tennant, 1988).

energy range near 14 meV, 30.5 meV, and 41.0 meV. The low-energy regime is explored in more detail in Fig. 3.15, which shows the transmission of $\lambda/2$ for a 5-cm-thick piece of PG in the 13–15 meV and 4.5–6.0 meV energy ranges and the transmission of $\lambda/3$ in the lower-energy range (Shapiro and Chesser, 1972). There are two minima in each range, with the most efficient occurring at 13.7 meV, where the transmission at λ is 78% (not shown) and at $\lambda/2$ is 0.01%. At 4.6 meV the transmission for $\lambda/2$ is 0.3% with an 80% transmission for λ (not shown). The $\lambda/2$ suppressions are due to the large cross section shown in Fig. 3.13 in the energy ranges of 60 and 20 meV, respectively. It should be noted that for the low-energy regime Be is a better filter, but its cutoff is 5.2 meV while PG can still be used effectively at 5.6 meV.

"Tuning" of the PG filter is crucial in that the beam has to be accurately aligned parallel to the c-axis in order for the filter to be effective. The mosaic width of filter-grade PG can be greater than the 0.5° required for monochromators. It was shown (Shirane and Minkiewicz, 1970) that a 2-cm-thick PG filter with a mosaic width of 3.5° has a transmission of 75%

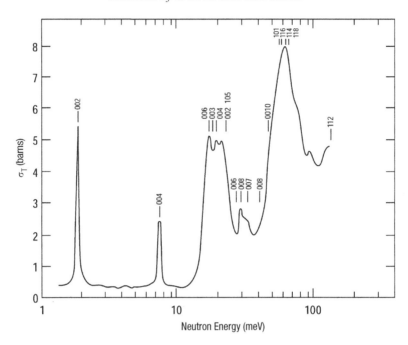

Fig. 3.13. Total cross section per atom of pyrolytic graphite (PG) as a function of energy for an incident neutron beam aligned along the *c*-axis (from Loopstra, 1966). The numbers correspond to scattering by various Bragg reflections.

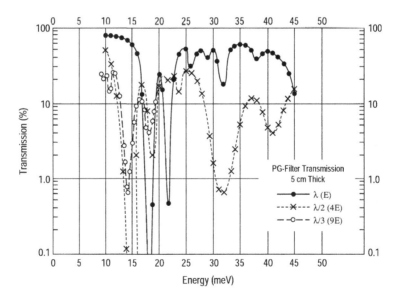

Fig. 3.14. Transmission of a 5-cm-thick PG filter as function of energy for desired energy E (wavelength λ), $4E$ ($\lambda/2$) and $9E$ ($\lambda/3$).

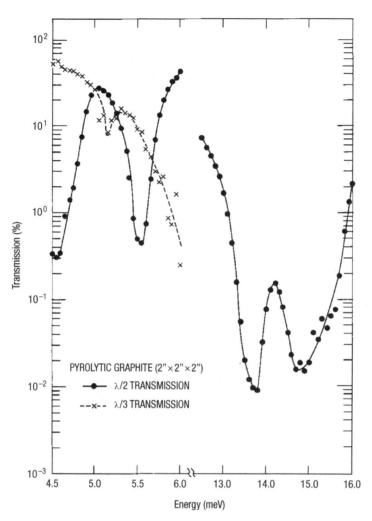

Fig. 3.15. Transmission of 5-cm-thick PG filter in the low-energy regime (from Shapiro and Chesser, 1972).

at 13.7 meV ($\lambda = 2.44\,\text{Å}$), and a transmission of $\sim 0.07\%$ for $\lambda/2$. With an increased mosaic width of $6.5°$, the transmission for λ decreases to 50% and $\lambda/2$ increases to 5%. A filter composed of higher-grade PG with a $1.0°$ mosaic width has a higher reduction level of $\lambda/2$ and a better transmission of λ, but the increased difficulties in tuning and the higher costs do not yet justify this improved performance.

The choice of filter length involves a balance between $\lambda/2$-suppression efficiency and attenuation of the desired signal at λ; typically a 5-cm-thick filter is a reasonable size. Even with a relative transmission of 10^{-3} for

$\lambda/2$ versus λ, it is often useful to be able to quickly add a second filter to the beam path to test whether an observed weak feature is real or is due to higher-order neutrons. If the intensity decreases substantially with the addition of the second filter, it is due to higher order. On the other hand, if the intensity decreases by less than a factor of two, the feature is not caused by higher order. (One must also consider the possibility of double-scattering; this is discussed in Chap. 7.)

3.5 Absorbers

The most commonly used absorber for neutron experiments is cadmium metal. Besides having a large absorption cross section, cadmium is a soft material that can easily be formed into convenient shapes to mask the beam. It is frequently used to attenuate unwanted scattering from components such as sample supports that intercept the neutron beam; however, there are two disadvantages of working with Cd. The neutron absorption leads to γ-ray fluorescence which can result in a radiation hazard for an experimenter working near a Cd-masked sample, as may occur during sample alignment. One useful alternative to Cd is paint containing Gd_2O_3; gadolinium is also a strong absorber. The second concern with Cd is that the absorption cross section decreases with increasing neutron energy; thus, masking that is effective at 5 meV may be insufficient at 80 meV.

Boron nitride is also used as a neutron absorber. In addition to its absorptive qualities, it can be machined into arbitrary shapes and will withstand high temperatures. It is, therefore, a nearly ideal material to use for mounting samples to be studied at high temperatures. Other materials that are useful for neutron shielding to reduce background and define beam paths are borated plastics and epoxies, which are commercially available. They can be molded or machined into any shape.

Finally, it is often necessary to attenuate the neutron beam when aligning the sample on strong Bragg peaks. Slabs of plastic or borated glass are useful and can be made of various thicknesses to attenuate the beam by varying factors. These can easily be calibrated, since it is often useful to know the true measured intensity.

3.6 Spectrometer alignment

Triple-axis spectrometers are, on the one hand, relatively simple instruments in concept while, on the other, they are extremely easy to misuse. To obtain accurate measurements, it is crucial to have a proper alignment of each component of the spectrometer. Misalignment can lead to reduced intensities and, even worse, to spurious results. These effects are caused by the strong correlations among the scattering geometries at the monochromator, sample and analyzer.

A simple, common mistake is illustrated here. An experimentalist begins an inelastic scattering measurement by *assuming* that the monochromator and analyzer are properly aligned. If this assumption is incorrect, because the previous user left the apparatus out of alignment (for example, the analyzer angle is mis-set by $0.5°$), it may be far from obvious. If the sample crystal is aligned for maximum intensity of a particular Bragg peak with the analyzer misaligned, the scattering angle $2\theta'$ and sample angle ϕ' will be different from the true values of 2θ and ϕ. These angles are mis-set in order to accommodate the analyzer's misadjustment. Even if one tries to readjust the analyzer after aligning the sample, the optimum intensity will always correspond to the same, wrong analyzer angle. The only way to correct this misalignment is to change to a two-axis mode (see below), optimize the sample orientation, and then place the analyzer in the Bragg position for elastic scattering.

3.6.1 Initial optical alignment

There are two distinct aspects to spectrometer alignment. The first is the mechanical alignment of the spectrometer itself, which should be performed only by experienced facility personnel before any experiments begin. The center lines of the Soller collimators α_1, α_2, and α_3 (see Fig. 3.8) must intersect the centers of rotation of the monochromator, sample, and analyzer. These alignments are best accomplished using an optical method. Once this alignment is complete, the experimentalist should never adjust the positioning of the collimators with respect to motions that are perpendicular to the beam direction.

3.6.2 Experimental alignment

Alignment of the three spectrometer axes by the experimentalist requires a systematic approach, in which components are adjusted one at a time in a manner chosen to avoid correlated errors. The specific steps are described below, with the aid of Figs. 3.1 and 3.8. It is assumed that an initial alignment has been performed optically, so that the centers of rotation of the three crystals are in line and the collimators are properly positioned. The determination of the true value of the angles and the wavelength can be performed with neutrons.

(i) Starting from the monochromator, one sets $2\theta_M$ near the desired angle, even though it is not known precisely. With nothing on the sample table, the sample and analyzer scattering angles, $2\theta_S$ and $2\theta_A$, are set at their nominal zero positions, so that the detector is looking directly at the beam from the monochromator. The analyzer crystal should either be rotated to an angle where it will not diffract, or else removed. A single horizontal collimator (defining the horizontal divergence) should be placed between the monochromator and the sample table. Collimation restricting the vertical divergence should be placed after the sample. With sufficient attenuation in place to avoid saturating the detector, the shutter is opened and the monochromator is rotated alternately about its vertical and horizontal axes (which are perpendicular to **G** of the monochromator). These rotations should be performed iteratively until a maximum of the intensity is obtained.

(ii) The next step is to set up the spectrometer in two-axis mode, with one (horizontal) collimator after the sample. A standard Q-independent scatterer, such as a thin cylinder of polycrystalline vanadium or a hydrogen-containing material, should be placed on the sample axis; with the detector placed at a finite angle, $2\theta_S \sim 2\theta_M$, an intensity measurement should be made with the beam attenuation removed. This intensity should be compared with an earlier measurement under the same experimental conditions. (Comparison with a regularly maintained record of standard intensities allows the user to readily ascertain whether the instrument is performing optimally. Deviations from historically reproducible intensities indicate a misalignment, which should be corrected before proceeding.) The translation of the monochromator is checked by measuring the scattering intensity as the vanadium sample is translated across the beam. The peak-intensity position marks the beam center and should coincide with the center of rotation of the sample axis; if they do not coincide, the monochromator should be translated to bring them into coincidence. Another way of checking that the center of the beam passes over the center of rotation is to use a half-mask, as shown in Fig. 3.16. The intensity scattered from the standard sample for the two possible orientations of the mask (differing by a 180° rotation of the mask) should agree with each other if the sample is properly centered.

(iii) Once the user is satisfied with the monochromator alignment, the next procedure is to simultaneously calibrate the incident wave vector and check for an offset in $2\theta_S$. This is performed using a polycrystalline sample with a well-known lattice spacing and at least three accessible

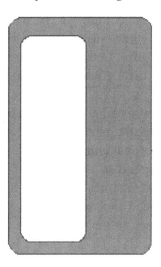

Fig. 3.16. Half-mask (approximately actual size) used to properly center a sample in the neutron beam at BNL.

Bragg peaks; Al_2O_3, Si, or Fe_3O_4 (magnetite) are possible candidates. Scans of three or more peaks are made, with the peak positions estimated from the nomimal value of k_i. From the measured angles θ_S^m for the reflections G_{hkl}, one performs a least-squares fit of the parameters k_i and $\Delta\theta_S$ to the formula

$$G_{hkl} = 2k_i \sin(\theta_S^m - \Delta\theta_S). \tag{3.26}$$

The offset or zero angle for $2\theta_S$ (required for converting from drive-motor units to calibrated angle) should be adjusted for $\Delta\theta_S$. From the newly determined k_i the correct values of θ_M and $2\theta_M$ are calculated from the formula

$$\theta_M = \sin^{-1}\left(\frac{2k_i}{G_M}\right). \tag{3.27}$$

(iv) The instrument is now aligned for two-axis (no analyzer) operation. The next procedure would be to place the sample in position and align it by using two orthogonal reflections. At each reflection, the tilt of the crystal is adjusted to give a maximum of the intensity. Translation of the sample to assure that it is centered in the beam can be tested with the half-mask (Fig. 3.16). Also, a transverse scan (rocking curve) and a longitudinal scan (θ-2θ) are performed at each of the two alignment reflections. The lattice parameter used to calculate $2\theta_S$ is adjusted so that the calculated $2\theta_S$ value agrees with the measured one. The sample-angle offset is adjusted so that the calculated sample angle corresponds with the maximum of the rocking curve. Note that the sample angle can only be adjusted for one reflection; after adjustment at one reflection, an offset between measured and calculated rocking-curve peaks for an orthogonal reflection indicates an error in the lattice parameters.

(v) The last step is to adjust the analyzer. This can be done using the beam diffracted from an aligned sample. First, the analyzer is rocked to find the position of maximum intensity, and the analyzer angle θ_A is adjusted to correspond to this position. Then the analyzer crystal and detector arm are scanned together in θ_A-$2\theta_A$ mode, and any necessary adjustment in the value of $2\theta_A$ is made. Alternatively, the analyzer can be aligned in the direct beam from the monochromator.

3.7 Goniometers

There are two types of sample goniometers used in neutron and X-ray scattering experiments: 1) a two-circle goniometer, which allows sample rotation about two orthogonal axes, and 2) a four-circle goniometer, which allows for sample rotation about three axes. The former is commonly used in triple-axis neutron instruments, while the latter is principally used in X-ray diffraction and elastic neutron crystallographic studies.

3.7.1 Two-circle goniometer

This type of goniometer is shown in Fig. 3.17(a). It consists of two tilt stages which allow rotations of the sample about two orthogonal axes, both of which are in the scattering plane. The radius of the tilt arcs should equal the distance from the beam center to the movable stays. For convenience, the tilt stages are placed on top of two orthogonal translational stages that allow for positioning of the sample at the center of the beam. This assembly is placed at the center of the sample rotation axis of a three-axis instrument. An advantage of this goniometer is that it can accommodate large and bulky ancillary equipment such as magnets, furnaces, cryostats, etc. A disadvantage is that one is generally restricted to measurements within a fixed plane of reciprocal space; motorized tilt axes allow scans over a small range out of that plane.

The two tilt stages are located one on top of the other and are referred to as the upper and lower arcs. It is important to realize that the two arcs are not equivalent, and the order in which they are adjusted in aligning a sample is crucial. Adjustment of the lower arc is independent of the tilt of

(a)

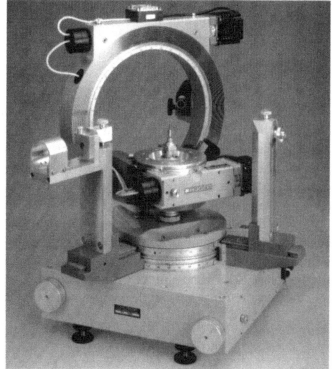

(b)

Fig. 3.17. (a) Photograph of a two-circle goniometer used in a triple-axis instrument. The upper tilt plate is 10 cm square. (b) Photograph of a four-circle goniometer used for neutron crystallography. The base plate is 36 cm square.

the upper arc, but the opposite situation is not true. When the upper arc is adjusted with the lower one already tilted, there is an additional component of rotation of the sample about a vertical axis. It is easier to visualize in the extreme limit of the lower arc being tilted by 90°. In this case, a tilt of the upper arc becomes a pure rotation of the sample about its vertical axis.

When aligning a sample to a particular scattering zone (with a particular reciprocal-lattice vector perpendicular to the scattering plane), the process involves using two orthogonal reflections, each aligned with one of the tilt axes. For example, suppose that one has an orthorhombic crystal, and is using the (200) and (020) reflections for alignment, with the (200) along the upper arc. The alignment procedure should begin with the spectrometer set for the (200) reflection. The upper arc is then adjusted to optimize the intensity. Next the spectrometer is moved to the (020) reflection, and the lower arc is adjusted to maximize the (020) intensity. After alignment, if one wishes to perform measurements involving tilting out of the selected scattering zone, it is important that only the lower arc be used for this purpose. This may require realignment of the sample with respect to the goniometer.

3.7.2 Four-circle goniometer

A four-circle goniometer, an example of which is shown in Fig. 3.17(b) allows three rotations for adjusting the sample orientation; these are the Eulerian angles θ, ϕ, and χ. The angle θ involves rotation about a vertical axis and corresponds to the usual sample angle θ_S. On top of the θ-drive is a circle standing on its edge; the χ angle involves motion about the center of this circle. Mounted on the vertical circle is the third drive, which provides the ϕ motion. With $\chi = 0$, the angles ϕ and θ become equivalent; when $\chi = 90°$, small motions of χ and ϕ are similar to the two-axis gonionmeter described above. By adjusting χ and ϕ, one can bring any plane of reciprocal space into the scattering plane. The disadvantage of this system is that weight and size limitations restrict the range of sample environments that can be used.

3.7.3 Translation of the sample

With a four-circle goniometer, two translation adjustments are provided above the ϕ rotation. In this manner the sample can always be brought into the exact center of the goniometer. Such positioning is essential for any type of X-ray scattering experiment where the beam size is small; if the sample is off-center, the apparent scattering angle, $2\theta_S$, will differ from the true value.

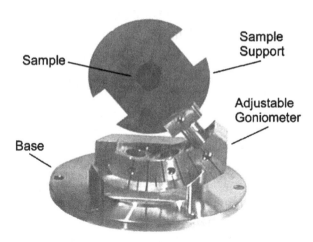

Fig. 3.18. Supergoniometer for adjusting sample orientations. The diameter of the base plate is 7.5 cm.

A similar arrangement is not practical for a neutron goniometer capable of handling large cryostats and superconducting magnets. Instead, the two translation stages are positioned below the two tilt stages; otherwise, the very large distance between the sample and the tilt axes would severely limit the tilt range. With this non-ideal arrangement, the sample will be at the center of the beam but not necessarily at the exact center of tilt rotations. With the larger beam and sample sizes typical of a neutron scattering experiment, the possible slight misalignment is generally not a significant problem.

3.7.4 Super holder

For neutron experiments involving a cryostat or a magnet, a single-crystal sample is placed in an aluminum can filled with He gas in order to provide good thermal contact with the cold finger and also to reduce any thermal gradients. Typically, it is very important that the crystal be nearly perfectly aligned within the can, especially when using a heavy superconducting magnet or a dilution refrigerator which cannot be tilted by more than a few degrees off vertical.

At Brookhaven, we have designed a special supergoniometer which is shown in Figure 3.18. The overall dimension of the base is 7.5 cm in diameter, large enough to avoid excess scattering by the surrounding can. In principle, it is similar to a small four-circle goniometer with a small χ circle and ϕ rotation which are adjusted by hand.

3.8 New developments

In recent years, significant efforts have been made to enhance the performance of triple-axis spectrometers. In particular, a number of enhancements were introduced simultaneously at Risø National Laboratory's TAS6 to create the "re-invented triple-axis spectrometer" or RITA (Mason *et al.*, 1995). Among the new features are: a supermirror guide within the biological shield to enhance the flux reaching a tall, vertically-focusing monochromator, a wide-band-pass neutron-velocity selector for elimination of higher-order neutrons, and a flexible analyzer–detector system allowing either horizontal focusing onto a single tube detector or a line-focus mode onto an area detector, with the latter providing a simultaneous measurement along a trajectory in $(\mathbf{Q}, \omega)$ space. The same analyzer–detector features have been implemented on the SPINS cold-neutron spectrometer at NCNR. For a review of other novel concepts, see Böni (2000).

References

Bacon, G. E. and Lowde, R. D. (1948). *Acta Cryst.* **1**, 303.

Bacon, G. E. (1975). *Neutron Diffraction* (Clarendon Press, Oxford), Chap. 3.

Böni, P. (2000). *Physica B* **276–278**, 6.

Bührer, W., Bührer, R., Isacson, A., Kock, M., and Thut, R. (1981). *Nucl. Instrum. Methods* **179**, 259.

Bührer, W. (1994). *Nucl. Instrum. Methods* **A338**, 44.

Carlile, C. J., Hey, P. D., and Mack, B. (1977). *J. Phys. E* **10**, 543.

Copley, J. R. D., Neumann, D. A., and Kamitakahara, W. A. (1995). *Can. J. Phys.* **73**, 763.

Currat, R. (1973). *Nucl. Instrum. Methods* **107**, 21.

Freund, A. (1976). *Conference on Neutron Scattering*. Report CONF-760601-p2, Oak Ridge National Laboratory, p. 1143.

Freund, A. and Forsyth, J. B. (1979). "Materials problems in neutron devices", in *Treatise on Materials Science and Technology, Vol. 15: Neutron Scattering*, ed. G. Kostorz (Academic Press, New York), p. 462.

Freund, A. K. (1983). *Nucl. Instrum. Methods* **213**, 495.

Hughes, D. J. and Harvey, J. A. (1955). *Neutron Cross Sections*, Report BNL-325, (Brookhaven National Lab).

Loopstra, D. O. (1966). *Nucl. Instrum. Methods* **44**, 181.

Magerl, A. and Wagner, V., eds. (1994). *Proceedings of Workshop on Focusing Bragg Optics, Nucl. Instrum. Methods* **A338**.

Mason, T. E., Clausen, K. N., Aeppli, G., McMorrow, D. F., and Kjems, J. K. (1995). *Can. J. Phys.* **73**, 697.

Meier-Leibnitz, H. (1967). *Ann. Acad. Sci. Fennicae, Phys. Ser.* VI, 267.

Mildner, D. F. R. and Lamaze, G. P. (1998). *J. Appl. Cryst.* **31**, 835.

Mughabghab, S. F., Divadeenam, M., and Holden, N. E. (1981). *Neutron Cross Sections* (Academic Press, New York).

Nieman, H. F., Tennant, D. C., and Dolling, G. (1980). *Rev. Sci. Instr.* **51**, 1299.

Nunes, A. C., and Shirane, G. (1971). *Nucl. Instrum. Methods* **95**, 445.

Popovici, M., Stoica, A. D., and Ioniță, I. (1987). *J. Appl. Cryst.* **20**, 90.

Pynn, R. and Passell, L. (1974). BNL Memo G-19.

Richter, D. and Shapiro, S. M. (1980). *Phys. Rev. B* **22**, 599.

Riste, T. and Otnes, K. (1969). *Nucl. Instrum. Methods* **75**, 197.

Riste, T. (1970). *Nucl. Instrum. Methods* **86**, 1.

Rustichelli, F. (1969). *Nucl. Instrum. Methods* **74**, 219.

Sailor, V. L., Foote, H. L., Jr, Landon, H. H., and Wood, R. E. (1956). *Rev. Sci. Instrum.* **27**, 26.

Scherm, R., Dolling, G., Ritter, R., Schedler, E., Teuchert, W., and Wagner, V. (1977). *Nucl. Instrum. Methods* **143**, 97.

Shapiro, S. M. and Chesser, N. J. (1972). *Nucl. Instrum. Methods* **101**, 1893.

Shirane, G. and Minkiewicz, V. J. (1970). *Nucl. Instrum. Methods* **89**, 109.

Soller, W. (1924). *Phys. Rev.* **24**, 158.

Tennant, D. C. (1988). *Rev. Sci. Instrum.* **59**, 380.

Vogt, T., Passell, L., Cheung, S., and Axe, J. D. (1994). *Nucl. Instrum. Methods* **A338**, 71.

4

Inelastic scattering and the resolution function

One of the chief strengths of the triple-axis spectrometer is its ability to measure the intensity of scattered neutrons for a particular momentum transfer $\mathbf{Q}_0$ and energy transfer $\hbar\omega_0$. This capability is especially important for measurements of inelastic scattering from single crystals. With a computer-controlled spectrometer, it is a simple matter to scan the energy transfer while sitting at a specific point in reciprocal space. Conversely, one may choose to scan the spectrometer along a particular direction in reciprocal space while maintaining a constant energy transfer. Such scans are easily interpreted, both qualitatively and quantitatively, as we will discuss.

Of course, because of the small scattering cross section for neutrons and the limited neutron flux generally available, one must typically perform measurements with finite beam divergences and with monochromator and analyzer crystals having significant mosaic widths. As a result, the energy and momentum transfers of the neutrons are distributed within some small region about the average values $(\omega_0, \mathbf{Q}_0)$. It is possible to describe the measured signal as a convolution of a spectrometer resolution function $R(\mathbf{Q}-\mathbf{Q}_0, \omega-\omega_0)$ and the scattering function $S(\mathbf{Q}, \omega)$. The resolution function is peaked at $(\omega_0, \mathbf{Q}_0)$ and decreases for deviations $(\Delta\omega, \Delta\mathbf{Q})$. The constant-amplitude contours for the resolution function form a set of nested ellipsoids in $(\omega, \mathbf{Q})$ space and, for a given spectrometer configuration, the volume, shape, and orientation of these ellipsoids depend only on $(\omega_0, \mathbf{Q}_0)$. The form of the measured spectra will depend on the way in which the resolution function is scanned through the structures defined by the scattering function. With some basic knowledge about the resolution function one can optimize scans to obtain sharper spectra and a better signal-to-noise ratio.

The first complete analysis of the shape of the resolution function for a triple-axis spectrometer was given by Cooper and Nathans (1967). The proper normalization of the resolution function was later described and

94

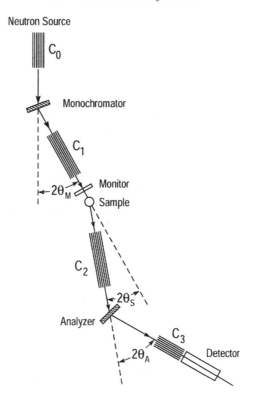

Fig. 4.1. Layout of a typical triple-axis spectrometer. C_j are collimators.

demonstrated by Chesser and Axe (1973). Much of the discussion presented here is based on these works. We will concentrate on factors that affect one's choice of spectrometer configuration and measurement scheme. When needed, a quantitative description of the resolution function can be obtained using commonly available computer programs. For those who are interested, mathematical details concerning the derivation and formulation of the resolution function are given in Appendix 4.

4.1 Notation and definitions

Figure 4.1 shows a schematic diagram of a typical triple-axis spectrometer. The angular divergence of the neutron beam along each leg of the beam path is typically limited by a collimator, C_j. It is generally assumed that the probability of transmission through a collimator as a function of divergence angle is a gaussian, with a width α_j in the horizontal (scattering) plane and β_j in the vertical direction. One generally specifies parameters α_j and β_j, the collimations, in terms of the full-width at half-maximum (FWHM) of the transmission function. Similarly, the monochromator crystal is assumed to have a gaussian mosaic distribution of width η_M in the horizontal plane and η_M' in the vertical; the analyzer crystal is characterized by mosaic widths η_A and η_A'.

The path of a scattered neutron in reciprocal space is illustrated in Fig. 4.2(a). The directions of the wave vectors $\mathbf{k}$ are the same as in real space, but the length of a wave vector is given by $2\pi/\lambda$, where λ is the neutron's wavelength. The average direction of the neutrons incident on the monochromator is determined by C_0, but their distribution in wavelength is quite broad compared to the range that can be diffracted by the monochromator. The neutrons incident on the sample, with wave vectors $\mathbf{k}_i$, are characterized by an average wave vector $\overline{\mathbf{k}}_i$ with length determined by the monochromator and direction selected by C_1. As discussed in Appendix 4, the distribution of the $\mathbf{k}_i$, denoted by $P_i(\mathbf{k}_i - \overline{\mathbf{k}}_i)$, can be calculated by taking the product of the transmission functions for the collimators C_0 and C_1 and the monochromator, and then integrating over the initial wave vectors $\mathbf{k}_i'$. Similarly, one can calculate the probability $P_f(\mathbf{k}_f - \overline{\mathbf{k}}_f)$ of detecting a scattered neutron with wave vector $\mathbf{k}_f$, where one integrates over all neutron wave vectors $\overline{\mathbf{k}}_f'$ that reach the dectector. Thus, the spectrometer defines distributions of incident wave vectors $\mathbf{k}_i$ that reach the sample and scattered wave vectors $\mathbf{k}_f$ that can reach the detector.

As we will discuss in the next section, the scattering properties of a sample depend on only the energy transfer $\hbar\omega$ and momentum transfer $\hbar\mathbf{Q}$, where

$$\hbar\omega = \frac{\hbar^2}{2m_n}(k_i^2 - k_f^2), \tag{4.1}$$

$$\mathbf{Q} = \mathbf{k}_f - \mathbf{k}_i. \tag{4.2}$$

The vector relationship for the momentum transfer is illustrated in Fig. 4.2(b). The average energy and momentum transfer, $\hbar\omega_0$ and $\hbar\mathbf{Q}_0$, are determined by the mean wave vectors $\overline{\mathbf{k}}_i$ and $\overline{\mathbf{k}}_f$. The shape of the resolution function is described in terms of the deviations of ω and $\mathbf{Q}$ from the average values. For the components of $\mathbf{Q}$ in the scattering plane, we define the deviations $\Delta\mathbf{Q}_\parallel$, parallel to $\mathbf{Q}_0$, and $\Delta\mathbf{Q}_\perp$, perpendicular to $\mathbf{Q}_0$, as indicated in Fig. 4.2. The vertical component, $\Delta\mathbf{Q}_z$, is perpendicular to the scattering plane. The coordinate system used is right-handed, and its relationship to the reciprocal lattice defined by the sample crystal is shown in Fig. 4.2(c).

The monochromator crystal of a spectrometer may scatter either to the right, as in Fig. 4.2(a), or to the left, as in Fig. 4.2(d). The left- and right-handed versions are related by a reflection through the vertical plane. Per-

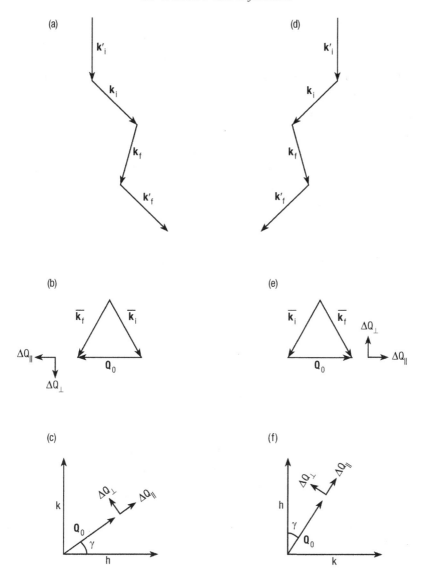

Fig. 4.2. (a) Reciprocal-space diagram of neutron path in a right-handed triple-axis spectrometer such as the one shown in Fig. 4.1. (b) Momentum-transfer diagram and relative coordinate system consistent with (a). (c) Relationship between **Q** and reciprocal lattice of the sample for a right-handed system, provided that the orientation of the sample is measured in the clockwise direction. (d)–(f) Same as (a)–(c), but for a left-handed spectrometer.

forming the same symmetry operation on the momentum-transfer diagram of Fig. 4.2(b), one obtains the result in Fig. 4.2(e). For a right-handed system, Q_0 will rotate in a counterclockwise direction with respect to the sample's reciprocal lattice [Fig. 4.2(c)] provided that the sense of rotation of the sample is measured in a clockwise direction. If one reverses all of the rotation angles in a left-handed system, then Q_0 will appear to rotate in a clockwise direction with respect to the sample. One can compensate for this by utilizing a left-handed coordinate system in reciprocal space. At Brookhaven, where the first triple-axis spectrometers were left-handed, it remained common practice to work with this left-handed coordinate system. While some of the examples presented elsewhere in this book involve a left-handed system, in this chapter we will follow standard convention and use the right-handed system of Fig. 4.2(b) and (c) (except where noted otherwise). All of the results discussed can be applied to the left-handed system by making the appropriate transformation.

One other sign convention can lead to confusion. As noted in Chap. 1, a frequently-used alternative to Eq. (4.2) is to set Q equal to $k_i - k_f$. Changing from one convention to the other has an important consequence for the orientation of the resolution ellipsoid in (ω, Q) space: it results in a reflection about $\omega = 0$. Experimentalists need to be aware of this fact when applying many of the results discussed in this chapter.

4.2 Definition of the resolution function

As discussed in the previous section, the monochromator, together with collimators C_0 and C_1, selects a bundle of neutrons with wave vectors k_i characterized by the distribution $P_i(k_i - k_i)$. The probability of a scattered neutron with wave vector k_f reaching the detector is determined by the distribution $P_f(k_f - \bar{k}_f)$. To calculate the neutron flux reaching the detector, we need to know the cross section for scattering an incident neutron with wave vector k_i into a differential volume of reciprocal space $d^3 k_f$ about k_f.

As discussed in Chap. 2, the differential scattering cross section is usually written in the form

$$\frac{d^2\sigma}{dE_f d\Omega_f} = \frac{k_f}{k_i} S(Q, \omega), \tag{4.3}$$

where $d\Omega_f$ is a differential element of solid angle, and

$$E_f = \frac{\hbar^2 k_f^2}{2m_n}. \tag{4.4}$$

For simplicity, we have assumed that factors such as the nuclear cross section are included in the scattering function; the important point here is that, except for the trivial k_f/k_i prefactor, the cross section depends only on Q and ω. Thus, Q and ω are the natural variables for describing a sample's scattering cross section, whereas $\mathbf{k}_i$ and $\mathbf{k}_f$ are the quantities defined by the spectrometer.

Consider a cartesian coordinate system defined relative to $\mathbf{k}_f$, with the z-axis along $\mathbf{k}_f$ and the x- and y-axes transverse to it. Then, differentiating Eq. (4.4) with respect to k_{fz} gives

$$dE_f = \frac{\hbar^2}{m_\mathrm{n}} k_f dk_{fz}. \tag{4.5}$$

For the solid angle we can write

$$d\Omega_f = \frac{dk_{fx} dk_{fy}}{k_f^2}. \tag{4.6}$$

Combining these results, we can write the differential cross section in the form

$$\frac{d^3\sigma}{dk_f^3} = \frac{\hbar^2}{m_\mathrm{n}} \cdot \frac{1}{k_f} \cdot \frac{d^2\sigma}{dE_f d\Omega_f} = \frac{\hbar^2}{m_\mathrm{n}} \cdot \frac{1}{k_i} S(\mathbf{Q}, \omega). \tag{4.7}$$

Combining the distributions P_i and P_f with the scattering cross section, we can write the flux reaching the detector as

$$F_\mathrm{d}(\bar{\mathbf{k}}_i, \bar{\mathbf{k}}_f) = \int d\mathbf{k}_i d\mathbf{k}_f \, F_i(k_i) P_i(\mathbf{k}_i - \bar{\mathbf{k}}_i) \frac{d^3\sigma}{dk_f^3} P_f(\mathbf{k}_f - \bar{\mathbf{k}}_f), \tag{4.8}$$

where $F_i(k_i)$ is the neutron flux at the first collimator C_0. The flux is equal to $k_i \phi(k_i)$, where $\phi(k)dk$ is the number of neutrons at C_0 with wave vectors in the range $\mathbf{k}$ to $\mathbf{k} + d\mathbf{k}$. The factor of k_i in the flux cancels the k_i in the differential cross section. We will assume that the variation of $\phi(k_i)$ is sufficiently slow compared to the narrow distribution defined by $P_i(\mathbf{k}_i - \bar{\mathbf{k}}_i)$ that we can replace k_i by its average value $\bar{k}_i$ and pull this quantity out of the integral. We now introduce the resolution function $R(\omega - \omega_0, \mathbf{Q} - \mathbf{Q}_0)$ which is defined by

$$F_\mathrm{d}(\omega_0, \mathbf{Q}_0) = \phi(\bar{k}_i) \int d\omega d\mathbf{Q} \, R(\omega - \omega_0, \mathbf{Q} - \mathbf{Q}_0) S(\mathbf{Q}, \omega), \tag{4.9}$$

where ω_0 and $\mathbf{Q}_0$ correspond to $\bar{\mathbf{k}}_i$ and $\bar{\mathbf{k}}_f$. It follows from Eqs. (4.7)–(4.9) that

$$R(\omega, \mathbf{Q}) = (\hbar^2/m_\mathrm{n}) \int d\mathbf{k}_i d\mathbf{k}_f \, P_i(\mathbf{k}_i) P_f(\mathbf{k}_f) \, \delta(\mathbf{Q} - \mathbf{k}_f + \mathbf{k}_i)$$
$$\times \delta[\omega - (\hbar/2m_\mathrm{n})(k_i^2 - k_f^2)]. \tag{4.10}$$

It is common practice to place a monitor between C_1 and the sample.

Assuming a monitor efficiency proportional to $1/k_i$, the flux measured by the monitor is

$$F_m(\bar{\mathbf{k}}_i) = \phi(\bar{k}_i) \int d\mathbf{k}_i \, P_i(\mathbf{k}_i - \bar{\mathbf{k}}_i) = \phi(\bar{k}_i)V_i, \qquad (4.11)$$

where

$$V_i = \int d\mathbf{k}_i \, P_i(\mathbf{k}_i - \bar{\mathbf{k}}_i). \qquad (4.12)$$

The detector signal is typically accumulated until a fixed number of monitor counts is detected; thus, the measured signal is

$$I(\omega_0, \mathbf{Q}_0) = F_d/F_m = \int d\omega d\mathbf{Q} \, R_{\mathrm{eff}}(\omega - \omega_0, \mathbf{Q} - \mathbf{Q}_0)S(\mathbf{Q}, \omega), \qquad (4.13)$$

where the effective resolution function is

$$R_{\mathrm{eff}} = R/V_i. \qquad (4.14)$$

Note that the resolution function as defined by Eq. (A4.3) depends in a symmetric fashion on $P_i(\mathbf{k}_i)$ and $P_f(\mathbf{k}_f)$, whereas, because of the normalization to the monitor signal, the effective resolution has lost that symmetry.

The resolution function is peaked at $\mathbf{Q}_0, \omega_0$ and falls off as $\mathbf{Q}$ and ω deviate from these values. Equation (4.9) says that what one actually measures is not $S(\mathbf{Q}, \omega)$ but rather a 4-dimensional convolution of $S(\mathbf{Q}, \omega)$ with the resolution function. In order to write down a functional form for $R(\omega - \omega_0, \mathbf{Q} - \mathbf{Q}_0)$ it is convenient to introduce the 4-vector $\mathcal{Q}$ such that

$$\mathcal{Q} = (\mathcal{Q}_0, \mathcal{Q}_1, \mathcal{Q}_2, \mathcal{Q}_3) = \left(\frac{m_n}{\hbar Q}\omega, Q_\parallel, Q_\perp, Q_z\right), \qquad (4.15)$$

where the components of $\mathbf{Q}$ are given in a coordinate system defined relative to $\mathbf{Q}_0$, with $Q_\parallel$ along the direction of $\mathbf{Q}_0$, $Q_\perp$ perpendicular and in the scattering plane, and Q_z perpendicular and in the vertical direction. If one uses the gaussian approximation (i.e., if one assumes that collimator transmission functions and the mosaic distributions of the monochromator and analyzer crystals are gaussians) it is possible to derive an analytic formula for the resolution function expressed as a 4-dimensional gaussian distribution (Cooper and Nathans, 1967)

$$R(\omega - \omega_0, \mathbf{Q} - \mathbf{Q}_0) = R_0 \exp\left(-\tfrac{1}{2}\Delta\mathcal{Q}M\Delta\mathcal{Q}\right), \qquad (4.16)$$

where

$$\Delta\mathcal{Q} \approx \left(\frac{m_n}{\hbar Q_0}(\omega - \omega_0), Q_\parallel - Q_0, Q_\perp, Q_z\right), \qquad (4.17)$$

and M is a 4×4 matrix (see Appendix 4 for details). Both R_0 and M are

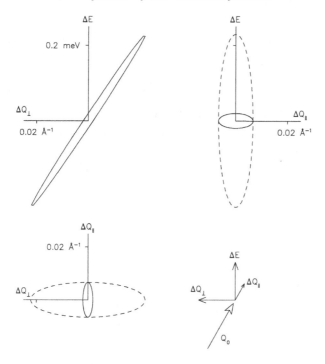

Fig. 4.3. Example of the resolution ellipsoid in $(\Delta E, \Delta Q_\|, \Delta Q_\perp)$ space for a right-handed spectrometer with monochromator and analyzer set to PG(002). Calculated for $\omega_0 = 0$, $Q_0 = 1.5\,\text{Å}^{-1}$, $E_i = 14.7\,\text{meV}$, horizontal collimations $(\alpha_0\text{-}\alpha_1\text{-}\alpha_2\text{-}\alpha_3)$ of $20'\text{-}20'\text{-}20'\text{-}20'$, and crystal mosaic widths of $\eta_M = \eta_A = 24'$. The solid lines represent the cross section of the ellipsoid with a given plane; dashed lines show the projection onto the plane.

functions of $\bar{k}_i$, $\bar{k}_f$, and $2\theta_S$. Setting the argument of the exponential in Eq. (A4.1) equal to a constant defines a 4-dimensional ellipsoid. In general, the resolution matrix M is not diagonal, and hence the principal axes of the resolution ellipsoid do not coincide with the axes defined by ω and $\mathbf{Q}_0$. The shape and orientation of the resolution ellipsoid will be discussed in §4.4. One simplification resulting from use of the paraxial approximation (i.e., the assumption that the beam divergence is small) is that resolution in the vertical $(\Delta \mathbf{Q}_z)$ direction is uncoupled from the other three coordinates. Hence the matrix M separates into a 3×3 matrix coupling ω, $\Delta Q_\|$, and $\Delta Q_\perp$, and a 1×1 matrix for ΔQ_z.

As an example, Figure A4.3 shows cross sections and projections of the three-dimensional ellipsoid for one set of spectrometer parameters. Because of the oblong, canted nature of the resolution ellipsoid, the effective resolution for a given experiment will depend not only on the spectrometer parameters, but also on the structure present in the scattering function. However, it should be noted that the resolution function depends only on the spectrometer configuration – it is completely independent of the sample itself. Before considering further details, it may be helpful to get a feeling for the magnitudes of the energy- and Q-widths for the resolution function. If we consider elastic scattering, with $E_i = E_f = E = \hbar^2 k^2 / 2m_n$, and take all of the horizontal collimator and mosaic widths to be equal to α, then $\Delta E / E \sim 2\alpha / \tan\theta_M$ and $\Delta Q / k \sim \alpha$. A collimator width α of $20'$ is equal to 0.00582 radians, or $\sim 0.6\%$. Using the (002) reflection of pyrolytic graphite for monochromator and analyzer and taking $E = 14.7\,\text{meV}$ gives $\Delta E \sim 0.5\,\text{meV}$ and $\Delta Q \sim 0.01\,\text{Å}^{-1}$.

4.3 Constant-Q scans

The standard form of inelastic measurement with a triple-axis spectrometer is the constant-**Q** scan, in which the momentum transfer is held fixed while the scattered intensity is measured as a function of the energy transfer. This type of scan, which was first introduced by Brockhouse (1961), is particularly convenient for measuring the dispersion of phonons and magnons. For ideal excitations (i.e., with no damping), $S(\mathbf{Q}, \omega)$ consists of surfaces defined by $\omega = \omega_j(\mathbf{Q})$, where j labels distinct dispersion branches. With **Q** fixed at $\mathbf{Q}_0$, measurements at a series of points ω_m will yield a peak when $\omega_m \approx \omega_j(\mathbf{Q}_0)$.

Besides mapping out dispersion curves, it is also possible to determine the magnitude of $S(\mathbf{Q}, \omega)$ associated with a point $\omega_j(\mathbf{Q}_0)$. Suppose, for simplicity, that the magnitude along the dispersion surface varies slowly in the region about $\omega_j(\mathbf{Q}_0)$ sampled by the resolution function. Then, assuming perfectly sharp excitations, we can write

$$S(\mathbf{Q}, \omega) \approx S_0 \delta[\omega - \omega(\mathbf{Q})], \tag{4.18}$$

in the vicinity of the measurement. Making use of Eqs. (4.13) and (4.14), the intensity measured at a point $(\omega_m, \mathbf{Q}_0)$ is then given by

$$I(\omega_m, \mathbf{Q}_0) = V_i^{-1} S_0 \int d\mathbf{Q} \, R[\omega(\mathbf{Q}) - \omega_m, \mathbf{Q} - \mathbf{Q}_0]. \tag{4.19}$$

In practice, one must measure $I(\omega_m, \mathbf{Q}_0)$ at a discrete set of points ω_m, and then sum the measured intensities to obtain the "integrated" intensity $\mathscr{I}_E$. For our theoretical example, we can integrate analytically. Making use of Eq. (A4.3), one finds (Dorner, 1972)

$$\int d\omega_m \int d\mathbf{Q} \, R[\omega(\mathbf{Q}) - \omega_m, \mathbf{Q} - \mathbf{Q}_0] = \int d(\Delta\mathbf{k}_i) \, P_i(\Delta\mathbf{k}_i) \int d(\Delta\mathbf{k}_f) \, P_f(\Delta\mathbf{k}_f)$$

$$= V_i V_f, \tag{4.20}$$

and hence

$$\int d\omega_m \, I(\omega_m, \mathbf{Q}_0) = V_f S_0, \tag{4.21}$$

independent of the form of the dispersion.

Equation (4.21) shows that the integrated intensity of a constant-$\mathbf{Q}$ scan can give a direct measure of $S(\mathbf{Q}, \omega)$, multiplied by the resolution volume associated with the analyzer arm of the spectrometer. If the final neutron energy, E_f, is held fixed, so that $\hbar\omega$ is varied by scanning E_i, then the phase-space volume V_f remains constant. (Keep in mind that the dependence on the monochromator arm is canceled out by measuring for the same number of monitor counts at each point.) Thus, measuring in the fixed-E_f mode provides a convenient way to map out the relative dependence of S as a function of $\mathbf{Q}$ for well-defined excitations $\omega(\mathbf{Q})$. (Determination of the absolute magnitude of S requires a normalization to the sample volume.)

Of course, there are times when it may be preferable to fix E_i and scan E_f. In that case, interpretation of the integrated intensities requires some knowledge of how V_f depends on various parameters. As shown in Appendix 4, V_f is given by

$$V_f = \frac{\bar{k}_f^3}{\tan \theta_A} R_A(\bar{k}_f) V'_f, \tag{4.22}$$

where $R_A(\bar{k}_f)$ is the reflectivity of the analyzer crystal, θ_A is the scattering angle at the analyzer, and V'_f is a constant that depends on collimator divergences and the mosaic width of the analyzer crystal. Chesser and Axe (1973) experimentally tested and verified some of the functional dependences in Eqs. (4.21) and (4.22) through a study of phonons in a single crystal of copper. They measured integrated intensities as functions of k_f and k_i. Some of their results will be discussed further in Chap. 5. Note that one can approximately cancel the $k_f^3 / \tan \theta_A$ dependence in fixed-E_i scans by having the data-collection program adjust the counting time (or number of monitor counts) at each point of the scan.

It is also of interest to consider the width of the peak measured in a constant-$\mathbf{Q}$ scan through a well-defined dispersion surface. To obtain analytic results, it is convenient to assume that the dispersion surface is locally planar, in which case we can write

$$\omega(\mathbf{Q}) = \omega(\mathbf{Q}_0) + \mathbf{c} \cdot \Delta\mathbf{Q}. \tag{4.23}$$

[For acoustic phonons, in the limit that $\mathbf{Q}_0 \to \mathbf{G}$ (a reciprocal-lattice vector), $\mathbf{c}$ would be the sound velocity.] Combining Eqs. (4.19), (4.23), and (A4.1), one can show that

$$I(\omega_m, \mathbf{Q}_0) = I_0 S_0 e^{-[\omega_m - \omega(\mathbf{Q}_0)]^2/2\Delta_\omega^2}, \qquad (4.24)$$

where

$$I_0 = V_f / \sqrt{2\pi}\Delta_\omega. \qquad (4.25)$$

Chesser and Axe (1973) derived a formula for the width Δ_ω in terms of the elements of the resolution matrix M of Eq. (A4.1). Their formula is quite useful for numerical calculations; however, the formula is rather complicated, and it is not easy to draw any useful conclusions by inspecting it. In general, the width depends on the orientation of the resolution ellipsoid with respect to the dispersion surface. The problem of matching the orientation of the resolution function to the dispersion is known as "focusing" and will be discussed in the next section.

Only in the limit of no dispersion ($\mathbf{c} = \mathbf{0}$) is it possible to derive a reasonably simple formula for the width. In that case one can show (Chesser and Axe, 1973; Nielsen and Bjerrum Møller, 1969) that

$$\hbar^2\Delta_\omega^2 = 4\left(E_i^2 B_i + E_f^2 B_f\right), \qquad (4.26)$$

where

$$B_i = \frac{1}{\tan^2\theta_M}\left(\frac{\alpha_0^2\alpha_1^2 + \alpha_0^2\eta_M^2 + \alpha_1^2\eta_M^2}{\alpha_0^2 + \alpha_1^2 + 4\eta_M^2}\right), \qquad (4.27)$$

and

$$B_f = \frac{1}{\tan^2\theta_A}\left(\frac{\alpha_2^2\alpha_3^2 + \alpha_2^2\eta_A^2 + \alpha_3^2\eta_A^2}{\alpha_2^2 + \alpha_3^2 + 4\eta_A^2}\right). \qquad (4.28)$$

These results can be applied to the energy width of a scan through the elastic incoherent scattering of a sample. Vanadium is especially useful for checking the energy resolution because of its particularly strong spin-incoherent scattering. Equations (4.26)–(4.28) simplify considerably if we restrict ourselves to elastic scattering, $E_i = E_f = E$, with all α_j and η_j equal to α, and $\theta_M = \theta_A$; in that case,

$$\frac{\hbar\Delta_\omega}{E} = \frac{2\alpha}{\tan\theta_M}. \qquad (4.29)$$

Table A3.3 in Appendix 3 lists values of $\hbar\Delta_\omega$ at various energies for a variety of monochromator choices, with $\alpha = 20'$.

4.4 Focusing

As mentioned earlier, momentum transfer in the vertical direction is de-coupled from the in-plane components and the energy transfer. Hence, in considering the shape and orientation of the resolution function we can restrict ourselves to a three-dimensional phase space. In general, the reso-lution ellipsoid has the shape of a flattened cigar (see Fig. A4.3). The peak width measured in a constant-Q scan through a sharp dispersion surface will depend on the orientation of the resolution function relative to the disper-sion surface. In a "focused" condition, the two longer axes of the resolution ellipsoid are parallel to the surface, and a narrow peak will be measured. At the opposite extreme, with the longest axis orthogonal, the peak may be so wide as to be undetectable. In the typical case, one does not strive for perfect focusing; instead, there is usually a choice between more focused and less focused measurement conditions. Here we will discuss the general rules for making this choice.

Consider a spectrometer in the conventional "W" configuration of Fig. 4.1, with a symmetric selection of collimations ($\alpha_0 = \alpha_3$, $\alpha_1 = \alpha_2$) and identical monochromator and analyzer crystals, as in the example of Fig. A4.3. Near $\omega = 0$ the resolution ellipsoid has a long axis in the $(Q_\perp, \omega)$ plane, with a negative slope. This orientation is favorable for measuring transverse acoustic phonons dispersing to the $-|\Delta Q_\perp|$ side of a Bragg peak in energy-loss mode (positive ω); an energy-gain measurement will be unfocused. A measurement on the $+|\Delta Q_\perp|$ side will be focused for energy gain instead of energy loss. As an example, Fig. 4.4 shows measurements of transverse acoustic phonons in MgO (Peckham, Saunderson, and Sharp, 1967). Clearly, finding the focused condition for a given measurement can be quite important.

Very close to $\omega = 0$ there is no focusing advantage for longitudinal acoustic modes; however, with increasing energy transfer the resolution ellipsoid rotates in the $(Q_\parallel, Q_\perp)$ plane, and the situation changes. Figure 4.5 shows some examples. Consider first the case of energy loss. For a given $\mathbf{Q}_0$, the orientation of $\mathbf{Q}_0$ becomes antiparallel to $\mathbf{k}_i$ and $\mathbf{k}_f$ at the maximum possible energy transfer. In this case, the resolution ellipsoid will tend to have a long axis near the $(Q_\parallel, \omega)$ plane with a positive slope $\Delta\omega/\Delta Q_\parallel$. For the case of energy gain, $\mathbf{Q}_0$ rotates to become parallel to $\mathbf{k}_i$ and $\mathbf{k}_f$ at the condition of greatest energy transfer. For this situation, the resolution ellipsoid has a long axis with a negative slope in the $(Q_\parallel, \omega)$ plane. At both extremes, the focusing condition corresponds to a positive dispersion.

Scans between energy-transfer extremes are not uncommon in small-angle (small-Q) measurements. For an amorphous or polycrystalline sample, the

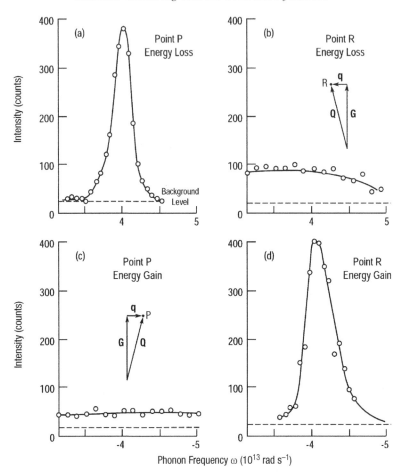

Fig. 4.4. Constant-**Q** scans of a transverse phonon propagating along a [111] direction in MgO measured for (a, b) energy gain and (c, d) energy loss at (a, d) focused and (b, c) unfocused positions using a right-handed spectrometer (from Peckham, Saunderson, and Sharp, 1967).

only region where one can study the dispersion of excitations is near $Q = 0$. Small-Q spin waves in a ferromagnet will increase in energy with increasing Q, so focusing effects will be similar to those for longitudinal acoustic phonons. An example of good focusing for both energy gain and energy loss is given in Fig. 4.6, which shows a measurement of spin waves in an amorphous metallic ferromagnet (Axe *et al.*, 1977).

It is simplest to conceptualize measurements of longitudinal or transverse excitations at positions **Q** displaced longitudinally or transversely from a Bragg point. However, because of the rotation of the resolution ellipse with

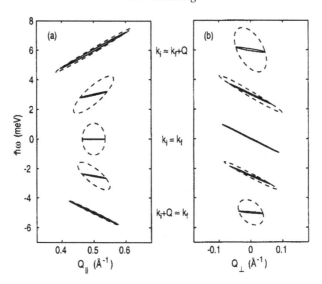

Fig. 4.5. Examples demonstrating focusing for $Q_\parallel$. Dashed line: cross section of the resolution ellipsoid in the (a) ω–$Q_\parallel$ plane, and (b) ω–$Q_\perp$; solid line: projection of the ellipsoid onto each plane. Calculated for $Q = 0.5\,\text{Å}^{-1}$, $E_f = 14.7\,\text{meV}$, horizontal collimations of 40′-40′-40′-40′, and $\eta_M = \eta_A = 24'$. Note that conditions for focusing in ω–$Q_\parallel$ are best at the extremes of energy transfer; the long axis of the resolution ellipsoid rotates as $\hbar\omega$ changes.

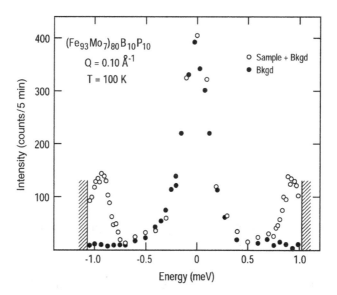

Fig. 4.6. Measurement of spin waves in an amorphous metallic ferromagnet at small **Q**, with $E_i = 14$ meV. The cross hatching indicates the limits of the scans. Notice that the widths of the spin-wave peaks are the same for energy gain and energy loss (from Axe *et al.*, 1977).

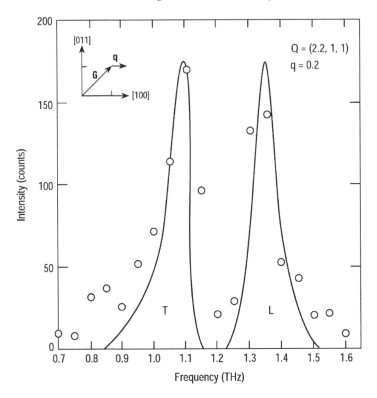

Fig. 4.7. A measurement of phonons propagating in the [100] direction in sodium metal. As indicated to the left, the scan was performed at a **q** having an intermediate orientation with respect to **G**. As a result, longitudinal and transverse modes are observed simultaneously, with comparable widths. The lines are the result of a calculation taking into account the resolution function and the curvature of the dispersion surface (from Werner and Pynn, 1971).

increasing energy transfer, the position for best focusing will not always be in the purely transverse direction, as it is for small ω. Narrower linewidths for both transverse and longitudinal acoustic modes may be obtained in scans at **Q** positions displaced in an intermediate direction. Figure 4.7 shows an example of such a scan measuring acoustic phonons in sodium metal (Werner and Pynn, 1971). In this case, the transverse and longitudinal modes have very similar linewidths.

So far we have discussed examples that all fall into the category of "gradient" focusing (Cooper and Nathans, 1967) in which the focusing effect depends on the slope of the longest axis of the resolution ellipsoid in the relevant (ω, Q_i) plane. An alternative form of focusing that is sometimes useful involves the orientation of the ellipsoid in the $(Q_\parallel, Q_\perp)$ plane. One

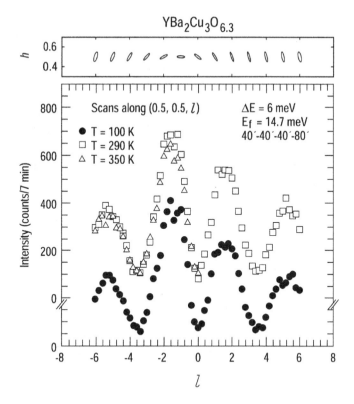

Fig. 4.8. An example of "Q" focusing. Constant-E scan along the 2D antiferromagnetic rod of scattering in the layered copper-oxide compound $YBa_2Cu_3O_{6.3}$. The sinusoidal oscillation, due to the structure factor, is symmetric in l. The enhancement near $l = -1.8$ is due to the focusing effect. The panel at the top illustrates how the projection of the resolution ellipsoid onto the (h, l) plane [where $Q = (h, h, l)$] rotates along the scan path, with optimal focusing at $l = -1$. Adapted from Tranquada *et al.* (1989).

such situation involves the measurement of steeply dispersing excitations in a two-dimensional (2D), or quasi-2D, system. If the dispersion is much greater than the $\Delta\omega/\Delta Q$ slope of the resolution function, then gradient focusing is not practical. For a 2D system, however, the scattering is independent of momentum transfers perpendicular ($q_\perp$) to the plane of the system. Thus, the steeply dispersing excitations essentially form a wall in the $(\omega, q_\perp)$ plane. A scan at fixed ω along $q_\perp$ will reach a maximum intensity when the longest axis of the resolution ellipsoid rotates into the $(\omega, q_\perp)$ plane. Such behavior has been observed in spin-wave measurements for the layered copper-oxide antiferromagnets. For example, Fig. 4.8 shows a scan along $(\frac{1}{2}, \frac{1}{2}, l)$ in antiferromagnetic $YBa_2Cu_3O_{6.3}$ (Tranquada *et al.*, 1989). (A left-handed coordinate system is used in the figure.) In this particular case there is a sinusoidal modulation along the $q_\perp$ direction, due to weak coupling between copper-oxide bilayers, which is symmetric in l. The large enhancement at $l \approx -1.8$ compared to $l \approx 1.8$ is due to "Q" focusing.

For elastic scattering, one is interested in the cross section of the resolution function with the $(Q_\parallel, Q_\perp)$ plane (as opposed to the projection considered above). The resolution widths at $\omega = 0$ can be measured by longitudinal and transverse elastic scans through a Bragg point. For a symmetric configuration, one can show that the general formulas for the longitudinal and transverse widths, $\Delta_\parallel$ and $\Delta_\perp$, are

$$\left(\frac{\Delta_\parallel}{Q_0}\right)^2 = \frac{a_{22} - 2a_{12}\cot\theta_S + a_{11}\cot^2\theta_S}{2(a_{11}a_{22} - a_{12}^2)} \tag{4.30}$$

and

$$\left(\frac{\Delta_\perp}{Q_0}\right)^2 = \frac{1}{2a_{22}}, \tag{4.31}$$

where the formulas for the matrix elements a are given in Appendix 4 [see Eqs. (A4.23)–(A4.26)]. The θ_S dependences of $\Delta_\parallel$ and $\Delta_\perp$ are illustrated in Fig. 4.9 for typical spectrometer parameters. $\Delta_\parallel$ has a minimum when $\theta_S \approx \theta_M = \theta_A$, and the minimum width will tend toward zero as $\eta_A, \eta_M \to 0$. In contrast, $\Delta_\perp/Q_0$ varies monotonically with θ_S, and $\Delta_\perp$ is typically much smaller than $\Delta_\parallel$. Since the transverse width scales with Q_0, it can be made quite small by reducing Q_0. An example is given by the work of Als-Nielsen and Laursen (1980), who took advantage of this fact to obtain extremely high-resolution measurements of critical scattering in the dipolar-coupled Ising ferromagnet $LiTbF_4$, as shown in Fig. 4.10.

4.5 Selection of collimators and energies

In the previous section we discussed how to optimize a measurement by selection of the measurement point in reciprocal space for a fixed spectrometer configuration. The choice of spectrometer parameters will also affect peak widths and intensities. In general, the choices involve a trade-off between intensity and resolution. Increasing the horizontal collimations (α_j) and k_i and k_f will rapidly increase both the intensity and resolution widths. The problem is to find the best compromise for a given problem, the so-called "experimental window".

Let us consider the situation close to the elastic condition. The effective volume of a single-axis resolution function depends symmetrically on the

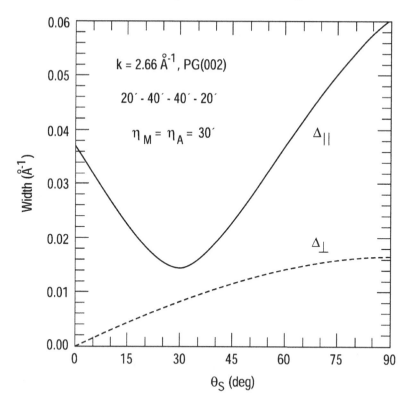

Fig. 4.9. Longitudinal and transverse widths for elastic scans through a single-crystal Bragg peak as a function of θ_S for typical spectrometer parameters.

horizontal collimations [see Eqs. (A4.38) and (A4.39) in Appendix 4]. For a symmetric spectrometer configuration, the slope of the long axis of the resolution ellipsoid in the $(Q_\perp, \omega)$ plane also has a symmetric dependence on α_0 and α_1, and on α_3 and α_2. When the slope of the resolution ellipse is not parallel to the dispersion surface being measured, the slope in the $(Q_\perp, \omega)$ plane will tend to limit the peak width measured in a constant-**Q** scan. It follows from the above considerations that there will be no particular advantage in decreasing one collimation versus another (e.g., α_0 relative to α_1) in terms of optimizing the intensity and effective resolution width. On the other hand, the longitudinal Q width $\Delta_\parallel$ has an asymmetric dependence on α_0 versus α_1 (and α_2 versus α_3). One can show that, for the same sacrifice in intensity, the resolution width for longitudinal elastic scans can be decreased much more by choosing $\alpha_0 < \alpha_1$ (and $\alpha_3 < \alpha_2$). Related considerations have been discussed by Kalus and Dorner (1973).

The incident and final neutron energies can also be adjusted to optimize focusing. Near the elastic condition one has $k_i \approx k_f \equiv k$. For a symmetric

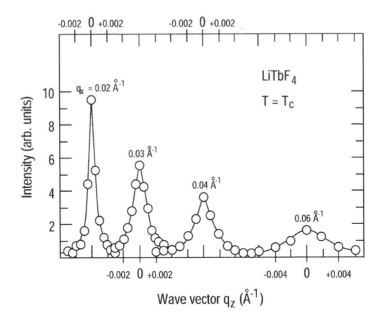

Fig. 4.10. Scans along the Ising axis q_z at different values of q_x at $T \approx T_c$ for the dipolar-coupled Ising ferromagnet LiTbF$_4$. The wave-vector dependence of the peak widths is an intrinsic property of the sample. Note the extremely high resolution that can be obtained for small momentum transfers (from Als-Nielsen and Laursen, 1980).

configuration with η_M and η_A much smaller than α_0 and α_1, one can show that the slope of the long axis of the resolution ellipse in the $(Q_\perp, \omega)$ plane is

$$\frac{\hbar \Delta \omega}{\Delta Q_\perp} \approx \frac{\hbar^2}{m_n} k \cos \theta_S [1 + \tan \theta_S \tan(\theta_S - \theta_M)],$$
$$\approx (4 \, \text{meV} \, \text{Å}^2) k, \tag{4.32}$$

where the second line applies for small θ_S. Thus, one can alter the value of $\Delta \omega / \Delta Q_\perp$ and attempt to match the velocity of acoustic excitations near zone center by varying k. In some cases, it may be more advantageous to increase k to match the velocity, rather than decreasing k to reduce the resolution volume. An example of such a case is shown in Fig. 4.11, where projections of the calculated resolution ellipsoid for different E_i are compared with the phonon dispersion curve for transverse acoustic phonons in niobium (Shapiro, Shirane, and Axe 1975). Using $E_i = 5 \, \text{meV}$ reduces the dimensions of the resolution ellipsoid; however, the peak width measured in a constant-$\mathbf{Q}$ scan will be little different at $E_i = 13.5 \, \text{meV}$ because of the improved

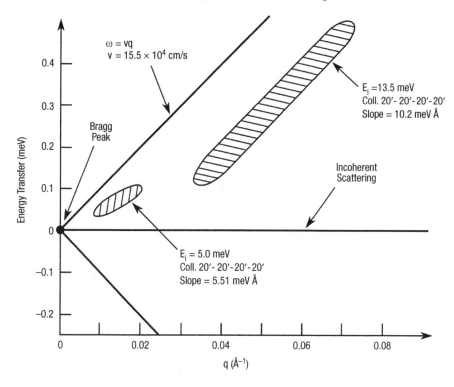

Fig. 4.11. Triple-axis spectrometer resolution function calculated for incident neutron energies of 13.5 and 5.0 meV, compared with the dispersion curve (solid lines) for transverse acoustic phonons in niobium (from Shapiro, Shirane, and Axe 1975).

match between the slope of the resolution function and the phonon velocity. Measurements at the higher energy will give much more intensity because of the larger resolution volume.

When the energy transfer becomes comparable to E_i and E_f, another criterion becomes important. The resolution volumes for the incident and final neutron beams are proportional to k_i^3 and k_f^3, respectively. The intensity will be limited by the side with smaller volume, while the resolution will tend to be limited by the side with the larger volume. A compromise between intensity and resolution can be achieved in this situation by changing to a non-symmetric configuration of collimators in order to balance the resolution volumes for $\mathbf{k}_i$ and $\mathbf{k}_f$. The ratio of resolution volumes is approximately

$$\frac{V_i}{V_f} \approx \left(\frac{k_i}{k_f}\right)^3 \frac{\alpha_0 \alpha_1 \tan \theta_A}{\alpha_2 \alpha_3 \tan \theta_M}, \tag{4.33}$$

where we have assumed identical monochromator and analyzer reflections, and that the monochromator and analyzer mosaic widths are comparable to the collimations. As an example, suppose $E_i = 30\,\text{meV}$, $E_f = 15\,\text{meV}$, and PG(002) reflections are used. Then collimations of 10′-20′-20′-20′ or 20′-40′-40′-40′ will make V_i/V_f approximately equal to unity.

Up to this point we have only considered optimizing the intrinsic scattering signal. Of course, in a real experiment there will always be background signal due to extrinsic sources such as the sample container and filters. The background signal can be reduced by tightening the collimation closest to the source. Thus, if background from the incident beam were a problem, one might tighten α_0 and α_1. Whether such a change improves the signal-to-background ratio will depend on the situation.

4.6 Bragg tails

To measure the resolution function for arbitrary conditions one would like to use a test sample with a scattering function of the form $\delta(\mathbf{Q} - \mathbf{Q}_0)\delta(\omega - \omega_0)$, where $\mathbf{Q}_0$ and ω_0 could be scanned through the resolution function. Since such a test sample does not exist, the next best thing is to scan the resolution function through a Bragg peak of a perfect crystal such as Ge, where $\mathbf{Q}_0$ is fixed and $\omega_0 = 0$. Provided that the $\mathbf{Q}$ dependence of the resolution matrix can be ignored, one can in this way map out the resolution ellipsoid.

The wings of a Bragg peak resulting from the oblong shape of the resolution function are often called "Bragg tails". While they can provide a useful measure of the resolution function, they are also an artifact which mimics zone-center inelastic excitations. As discussed in §4.4, the resolution function for elastic scattering (and a symmetric spectrometer configuration) has a major (long) axis in the $(Q_\perp, \omega)$ plane. For small energy transfers this orientation will change very little; thus, for a constant-$\mathbf{Q}$ scan transverse to a Bragg peak, a scattering peak will be observed at a finite energy transfer when the tail of the resolution function sweeps through the Bragg-peak position. The approximate dispersion of this feature, which looks like a transverse acoustic phonon branch, is given by Eq. (4.32). The Bragg tail can easily be identified because (i) as $\mathbf{Q}$ approaches the Bragg peak the intensity increases exponentially (in contrast to the power-law variation of phonon and magnon intensities), and (ii) the dispersion depends on k. It is a good practice when working at low energies near to a Bragg peak to perform a few test scans in order to identify the Bragg tail so as to prevent any confusion. For a crystal with a large mosaic width, in particular, the Bragg tail can become a cause of confusion.

4.7 Constant-E scans

In §4.3 we considered constant-Q scans for an idealized sample with a scattering function given by Eq. (4.18). For a planar dispersion surface the measured intensity is given by Eq. (4.24). The qualitative behavior of the peak width Δ as a function of the dispersion slope is sketched in Fig. 4.12. For small excitation velocities, $c = d\omega/dq$, the peak width will reach a minimum at the point of maximum focusing. As the velocity increases, the width will gradually increase, eventually diverging for infinite velocity. When the constant-Q peak width becomes large, it may be advantageous to hold the energy transfer fixed and instead scan Q. As indicated in the figure, the peak width for a constant-E scan will diverge for small velocities, but it will saturate at a finite value for large velocities.

For constant-E scans, both k_i and k_f are held fixed, so that the k_f dependence of the integrated intensities is a constant, and the measured intensities should be proportional to $S(Q,\omega)$. However, there is one other important difference from constant-Q scans (Brockhouse, 1964). The integrated intensity (for planar dispersion) is obtained by integrating Eq. (4.24) over Q_0 instead of ω_0. To perform the integral over the gaussian, one must transform the integration variable from Q_0 to ω_0, which requires multiplying the integral by the Jacobian J of the transformation, given by

$$J = 1/|d\omega/dq| = 1/c. \qquad (4.34)$$

Hence, integrated intensities measured in constant-E mode will depend inversely on the excitation velocity. This situation is indicated at the bottom of Fig. 4.12. Ignorance of this correction can lead to significant errors in the interpretation of constant-E measurements.

4.8 Vertical resolution

In discussing focusing we ignored the component of the resolution function for momentum transfers out of the scattering plane because, under the assumption of small beam divergence, it is decoupled from the in-plane components and from the energy transfer. While one typically ignores it in considering the best spectrometer configuration, finite vertical resolution can have important consequences for inelastic measurements.

To gain intensity, it is common practice to work with a relaxed vertical collimation. As a result, the resolution function is much broader for momentum transfers perpendicular to, rather than in, the scattering plane. While poor vertical resolution typically has no qualitative effect on measurements, there

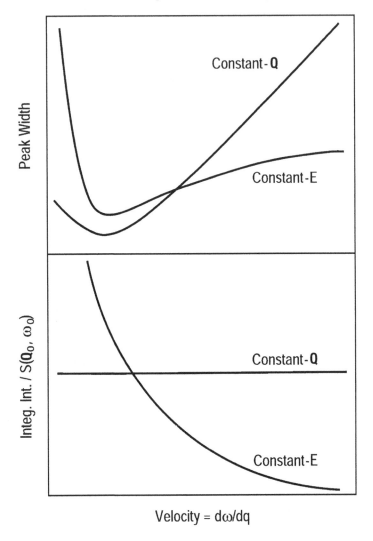

Fig. 4.12. Qualitative behavior of the peak width and integrated intensity as a function of excitation velocity for constant-**Q** and constant-*E* measurements.

are cases in which it can lead to the observation of unexpected "spurious" peaks. An important example involves phonon measurements at small **q**.

The scattering cross section for phonons contains the factor $(\mathbf{Q} \cdot \xi)^2$, where ξ, the phonon eigenvector, points in the direction of the atomic displacements. Along directions of high symmetry in a crystal, the eigenvectors correspond to purely longitudinal and transverse displacements. Thus, by restricting measurements to symmetry directions, one can make use of the eigenvector factor to selectively measure either longitudinal or transverse modes. However, this selectivity can be violated if two conditions are simultaneously satisfied: (i) the longitudinal and transverse modes are nearly degenerate, and (ii) the eigenvectors for the "forbidden" modes change their orientation within the resolution volume of the spectrometer. Both of these conditions are most easily satisfied for acoustic phonons at small **q**, and the second condition usually involves momentum transfers in the vertical direction.

The spurious peak problem is illustrated in Fig. 4.13, which shows constant-**Q** scans of acoustic phonons at small **q** in a Cu single crystal (Skalyo and Lurie, 1973). The measurements were taken in a [0$\bar{1}$1] scattering zone near the (200) Bragg peak. While the scan of the [110] transverse mode is clean, the scans of [100] longitudinal phonons are contaminated by scattering from transverse modes, the transverse contribution becoming dominant at small q. This problem can occur away from small **Q** in special cases. The observation of a spurious phonon branch due to transverse modes in measurements of [001] longitudinal phonons in alkali metals has been discussed by Copley (1971) and by Werner and Pynn (1971).

A less dramatic effect due to poor vertical resolution occurs when the dispersion of the excitations under study is very large in the vertical direction. In that case, the resolution function averages over a large range of excitation energy, resulting in a poor effective energy resolution. For example, in the layered antiferromagnet La_2CuO_4 the Cu-spins are strongly coupled within two-dimensional sheets, but weakly coupled between sheets. Spin waves in the planes have an extremely large velocity, but there is negligible dispersion perpendicular to the planes (Birgeneau and Shirane, 1989). When the perpendicular direction lies in the scattering plane, then one of the strong dispersion directions is vertical, and the effective resolution is poor. However, if a crystal is aligned to have its CuO_2 sheets in the scattering plane, then the broad vertical resolution is aligned with the weak dispersion direction, and it becomes possible (at sufficiently high energy transfers) to resolve the spin-wave peaks (Aeppli *et al.*, 1989).

4.9 Harmonics in the incident beam

A complication in the analysis of intensities measured in constant-**Q**, E_f-fixed mode arises from the presence of harmonics in the incident beam. The monochromator crystal will, in general, transmit not only neutrons at the fundamental energy E_1 but also at the harmonic energies $E_n = n^2 E_1$, where $n \geq 2$. Neutrons at unwanted energies can be eliminated from the scattered beam by placing a filter after the sample. The problem occurs with

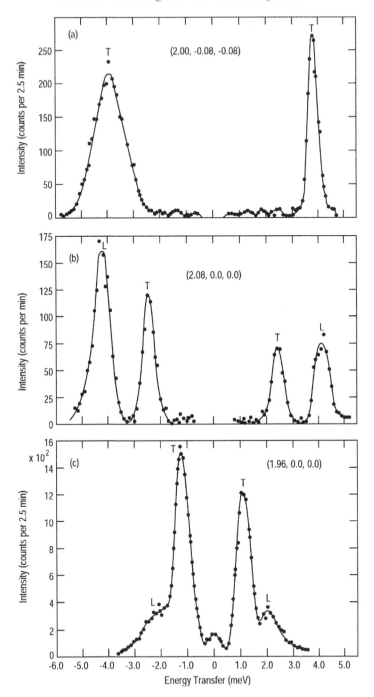

Fig. 4.13. Three typical phonon scans in copper. (a) Measurement of [110]T mode where only one phonon branch is observed for both phonon annihilation and creation. (b) Nominal measurement of [100]L mode. The peaks marked T are due to the large vertical resolution. (c) Nominal measurement of [100]L mode where the desired longitudinal response is obscured (from Skalyo and Lurie, 1973).

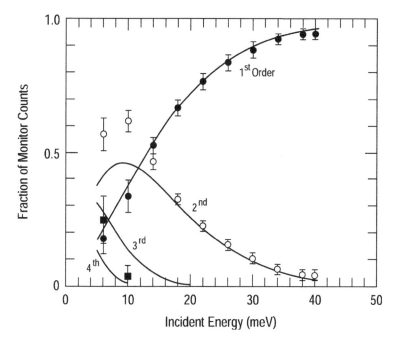

Fig. 4.14. Harmonic content versus incident energy. Points indicate measured values
(• first order, ○ second order, ■ third order) and lines are calculated, as discussed in
text.

the incident-beam monitor. As discussed in §4.1, one normally integrates
the scattered signal at each incident energy for a fixed number of monitor
counts, assuming that the monitor sees only fundamental neutrons. In reality,
the monitor will detect the fundamental plus the harmonics. If the harmonic
content changes significantly as the fundamental is varied, the scattered
signal will not be properly normalized. One solution to this problem is to
eliminate the harmonics by a judicious choice of monochromator crystal. For
example, at Chalk River a strained Si (111) crystal is often used; the second
harmonic is forbidden for this orientation. An alternative is to measure the
harmonic content, and then correct the intensities appropriately.

The latter approach has been taken at most laboratories. Figure 4.14 shows
measurements from the HFBR at Brookhaven of the harmonic content
of the monitor signal as a function of incident energy measured with a
series of attenuators (Patterson, Shirane, and Cowley 1976). The fraction of
the monitor signal due to harmonics increases rapidly as the fundamental
decreases below the peak in the incident flux. The measurements can be
fit with a simple model. Neutrons coming from the reactor have been
thermalized by the moderator, and are assumed to have velocities which

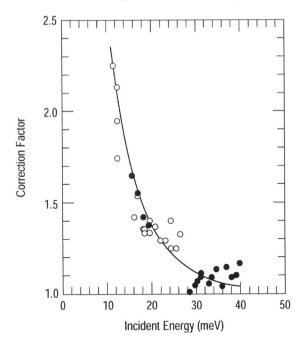

Fig. 4.15. Correction factor for fixed-E_f scans versus incident neutron energy. Points indicate measured values (• $E_f = 14.8$ meV, ○ $E_f = 24.0$ meV) and the solid line is calculated.

follow the Maxwell distribution law. The number of neutrons dn with wave vectors in the range $\mathbf{k}$ to $\mathbf{k} + d\mathbf{k}$ is then

$$dn = \phi_0 \left(\frac{\hbar^2}{2\pi m_n k_B T}\right)^{3/2} \exp\left(\frac{-\hbar^2 k^2}{2m_n k_B T}\right) d\mathbf{k} = \phi(k)d\mathbf{k}, \qquad (4.35)$$

where ϕ_0 is a constant and T is the temperature of the moderator. The differential neutron flux in the direction of $\mathbf{k}$ is proportional to kdn, and we can write

$$kdn \sim k^3 \exp(-\hbar^2 k^2/2m_n k_B T)dk. \qquad (4.36)$$

Taking into account the $1/k$ monitor sensitivity (and ignoring the weak energy dependence of the monochromator reflectivity), the total monitor signal should be given by

$$M_T \sim CE_1 e^{-E_1/k_B T} \qquad (4.37)$$

where

$$C = \sum_{n=1}^{\infty} n^2 e^{-(n^2-1)E_1/k_B T}. \qquad (4.38)$$

The fraction of the monitor signal due to the nth harmonic is

$$R_n = C^{-1} n^2 e^{-(n^2-1)E_1/k_B T}. \tag{4.39}$$

The solid curves in Fig. 4.14 were calculated using $k_B T = 25\,\text{meV}$. (Note that the effective source temperature at other reactors may be different.) The measured intensities for the scattered beam can be corrected by multiplying by C. In Fig. 4.15 the calculated correction factor C is compared with phonon data measured for the $[110]T_2$ acoustic branch in a copper single crystal (Patterson, Shirane, and Cowley, 1976). The points in the figure represent the ratio of measured integrated intensities to theoretical intensities. The two sets of measurements, with $E_f = 14.8$ and $24\,\text{meV}$, were normalized to match the theoretical curve. Below $\sim 10\,\text{meV}$ the correction factor becomes so large that quantitative analysis of intensities becomes difficult.

Sometimes it is more convenient to work with E_i fixed. For example, at small incident energies harmonics can become overwhelming; using a Be filter in the incident beam, one can eliminate this problem. Of course, when analyzing intensities, one must take into account the energy dependence of the analyzer reflectivity, and also the strong k_f dependence of the resolution volume, as discussed in § 4.3.

References

Aeppli, G., Hayden, S. M., Mook, H. A., Fisk, Z., Cheong, S.-W., Rytz, D., Remeika, J. P., Espinosa, G. P., and Cooper, A. S. (1989). *Phys. Rev. Lett.* **62**, 2052.

Als-Nielsen, J. and Laursen, I. (1980). In *Ordering in Strongly Fluctuating Condensed Matter Systems*, ed. T. Riste (Plenum, New York), p. 39.

Axe, J. D., Shirane, G., Mizoguchi, T., and Yamauchi, K. (1977). *Phys. Rev. B* **15**, 2763.

Birgeneau, R. J., and Shirane, G. (1989). In *Physical Properties of High Temperature Superconductors*, ed. D. M. Ginsberg, (World Scientific, Singapore), pp. 151–211.

Brockhouse, B. N. (1961). In *Inelastic Scattering of Neutrons in Solids and Liquids*, (IAEA, Vienna), p. 113.

Brockhouse, B. N. (1964). In *Phonons and Phonon Interactions* (ed. T. A. Bak) (W. A. Benjamin, Inc., New York) p. 221.

Chesser, N. J. and Axe, J. D. (1973). *Acta Cryst. A* **29**, 160.

Cooper, M. J. and Nathans, R. (1967). *Acta Cryst.* **23**, 357.

Copley, J. R. D. (1971). *Solid State Commun.* **9**, 531.

Dorner, B. (1972). *Acta Cryst. A* **28**, 319.

Kalus, J. and Dorner, B. (1973). *Acta Cryst. A* **29**, 526.

Nielsen, M. and Bjerrum Møller, H. (1969). *Acta Cryst. A* **25**, 547.

Patterson, H., Shirane, G. and Cowley, R. A. (1976). BNL Memo G-29.

Peckham, G. E., Saunderson, D. H., and Sharp, R. I. (1967). *Brit. J. Appl. Phys.* **18**, 473.

Shapiro, S. M., Shirane, G., and Axe, J. D. (1975). *Phys. Rev. B* **12**, 4899.

Skalyo, Jr., J. and Lurie, N. A. (1973). *Nucl. Instrum. Methods* **112**, 571.
Tranquada, J. M., Shirane, G., Keimer, B., Shamoto, S., and Sato, M. (1989). *Phys. Rev. B* **40**, 4503.
Werner, S. A. and Pynn, R. (1971). *J. Appl. Phys.* **42**, 4736.

5

Phonons and magnons

So far we have described how a triple-axis spectrometer works, and presented some of the basic techniques of inelastic neutron scattering. In this chapter we will illustrate some of these ideas through a discussion of several simple examples. We begin by considering acoustic phonons in face-centred-cubic (fcc) Cu and how to measure them. The next system studied is the prototypical ferromagnet body-centered-cubic (bcc) Fe, for which both phonons and magnons are of interest. Finally, we discuss the measurement of spin waves in an antiferromagnetic system, MnF_2. These topics are illustrated with results from the literature, complemented by some measurements that we have made for demonstration purposes.

5.1 Phonons in copper

Copper has an fcc lattice with:

$$a = 3.615 \text{ Å}, \qquad \text{space group} = Fm3m.$$

Although we tend to think of the fcc structure in terms of a simple cubic lattice with a four-atom basis, it is actually a Bravais lattice with a single atom per primitive unit cell. As a result, there are only three phonon modes (one per spatial dimension) at each wave-vector $\mathbf{Q}$ in the Brillouin zone. Considering again the conventional cubic cell description, the reciprocal lattice is also cubic, with unit-cell length $a^* = 2\pi/a$. Because of the four-atom basis, there is a structure factor for each reciprocal-lattice point, with a value of either 4 or 0. The points with finite structure factor are given by the fcc extinction rule

$$hkl \text{ all even or all odd.}$$

Because of the small number of phonon branches, together with the large

Cu Phonons

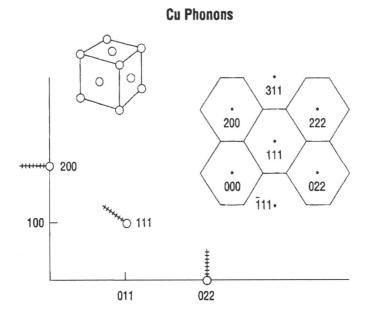

Fig. 5.1. Diagram of the (*hkk*) zone of reciprocal space for fcc Cu. Hatched lines indicate locations for measurement of low-energy transverse acoustic phonons propagating along high-symmetry directions with optimal focusing. *Left inset*: conventional unit cell for the fcc structure. *Right inset*: lines indicate the first Brillouin zone about several reciprocal-lattice points in the (*hkk*) zone.

coherent scattering length for Cu ($b = 0.76 \times 10^{-12}$ cm), the net scattering cross section for phonons in Cu is quite large. The availability of large, high-quality single crystals makes Cu a useful standard for inelastic measurements. The only drawback is that the large value of a^* (1.738 Å^{-1}) limits the number of Brillouin zones accessible with the most commonly used neutron energy, 14.7 meV.

The fcc unit cell and the first few Brillouin zones in the (*hkk*) scattering zone are depicted in the insets of Fig. 5.1. Detailed measurements and analyses of the phonon dispersions for Cu along directions of high symmetry have been reported by Svensson, Brockhouse, and Rowe (1967) and by Nicklow *et al.* (1967). Some of these results are shown in Fig. 5.2. To convert from frequency units, as given in the figure, to energy units, the conversion factor is

$$1 \text{ THz (terahertz)} = 10^{12} \text{ cps (counts s}^{-1}) = 4.13 \text{ meV}.$$

The hatched lines in Fig. 5.1 indicate regions where low-energy acoustic modes can be measured along symmetry directions with optimal focusing.

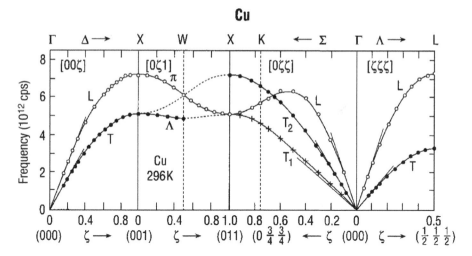

Fig. 5.2. Phonon dispersion curves for fcc Cu at 296 K as measured by Svensson, Brockhouse, and Rowe (1967).

The energy-integrated intensity for phonon creation measured in a constant-**Q** scan is given by

$$\mathcal{I}_E \sim F(k_f)[n(\omega) + 1]\frac{|\mathbf{Q} \cdot \boldsymbol{\xi}_j|^2}{\hbar\omega}e^{-2W(\mathbf{Q})}, \tag{5.1}$$

where $\boldsymbol{\xi}_j$ is the eigenvector of the jth phonon branch, and $\exp[-2W(\mathbf{Q})]$ is a Debye–Waller factor. The factor $F(k_f)$ is given by

$$F(k_f) = R_A(k_f)k_f^3/\tan\theta_A, \tag{5.2}$$

where R_A is the reflectivity of the analyzer and θ_A is the scattering angle of the analyzer. This factor accounts for changes in analyzer response and resolution volume as a function of k_f. The relevance of $F(k_f)$ for fixed-E_i and fixed-E_f scanning modes will be discussed below.

For a given phonon energy, the variation of the integrated intensity in reciprocal space is determined largely by the factor $|\mathbf{Q} \cdot \boldsymbol{\xi}_j|^2$. For Cu, with one atom per primitive unit cell, one has $|\boldsymbol{\xi}_j| = 1$. Along directions of high symmetry in reciprocal space the eigenvectors will point either parallel or perpendicular to the propagation wave vector. Thus, one can often pick a measurement geometry such that $\mathbf{Q} \parallel \boldsymbol{\xi}_j$, so that

$$\mathcal{I}_E \sim Q^2. \tag{5.3}$$

This relation was experimentally verified for longitudinal acoustic (LA) phonons in Cu by Chesser and Axe (1973) as shown in Fig. 5.3. Besides

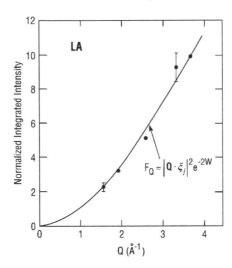

Fig. 5.3. Integrated intensities as a function of Q measured in E_i-fixed mode, corrected for $\hbar\omega$ and normalized at 3.6 Å^{-1}. Only results for longitudinal acoustic (LA) phonons are included (from Chesser and Axe, 1973).

correcting the integrated intensities for the phonon energy, they also included the Debye–Waller factor in the curve of calculated intensity, using the experimentally determined result

$$W(Q) = (0.01) \times (Q/a^*)^2. \tag{5.4}$$

The Q^2 dependence of the integrated intensity shows that it is advantageous to measure phonons at large Q. However, in selecting scattering conditions for an arbitrary case, one must take into account the possible adverse effects of a finite crystal mosaic width as well as the Debye–Waller factor at large Q values. Also, for more complicated crystals, the magnitudes of the eigenvectors may have a substantial Q dependence that does not have the periodicity of the reciprocal lattice. In such a case, the eigenvector magnitudes of acoustic modes in a given Brillouin zone will become equal to the static structure factor as $q \to 0$. Thus, Bragg intensities provide a useful guide to the $|\mathbf{Q} \cdot \boldsymbol{\xi}_j|^2$ factors for acoustic modes.

Now let us discuss the effects of focusing for measurements of transverse acoustic (TA) phonons. As an example, we will consider phonons measured near (200) with $\mathbf{q} \parallel [011]$. These phonons correspond to the T_2 branch in Fig. 5.2, and the slope of the dispersion at small q (velocity of sound) is identical to that of the [100]T mode in cubic crystals. A summary of typical measurements is shown in Fig. 5.4, and two representative phonon scans are presented in Fig. 5.5. The measurements were performed

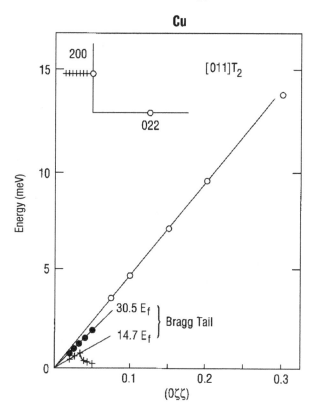

Fig. 5.4. Results of typical measurements of [011]T_2 phonons in Cu. The open circles indicate the observed phonon energies; filled circles and crosses indicate positions of peaks due to the Bragg tail. *Inset:* hatched line indicates locations of constant-**Q** scans in reciprocal space.

with fixed final energies of 14.7 and 30.5 meV, which correspond to the energy windows of the pyrolytic graphite filter commonly used in this energy range.

For measurements at very small q, close to the Bragg peak, the phonon signal is obscured by the much stronger Bragg-tail scattering, as discussed in §4.6. The slope of the Bragg tail is the same as the slope of the long axis of the resolution function (see §4.5), which is given approximately by

$$\frac{dE}{dq} \approx (4 \text{ meV Å}^2) \cdot k. \tag{5.5}$$

The Bragg tail can easily be mapped out by quick scans in the small-q range. (In fact, it is good practice to characterize the Bragg tail at the start of a phonon study, in order to avoid confusion caused by accidental measurement

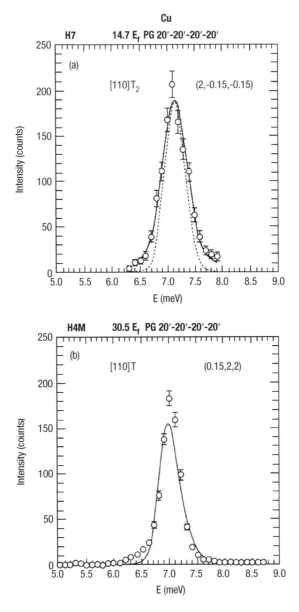

Fig. 5.5. Typical constant-**Q** phonon scans. Dashed line in (a) indicates the effective resolution width. The solid lines are fits to a lorentzian lineshape convolved with the resolution function.

of such peaks.) Since the Bragg tail is a measure of the resolution function, it provides a useful indication of the degree of focusing (i.e., how well the slope of the resolution function matches the phonon dispersion). As one can see from the data in Fig. 5.4, the focusing is better for 30.5 meV than for

14.7 meV. The focusing effect results in a sharper phonon peak profile (see Fig. 5.5) for 30.5 meV even though the resolution ellipsoid is considerably larger than for 14.7 meV. The measured Bragg tail at 14.7 meV deviates from a straight line at larger q. Such behavior is frequently observed; it is probably due to non-gaussian features of the sample mosaic distribution or resolution function.

5.1.1 E_i-fixed mode

For quantitative analysis of integrated intensities, it is important to correct for the $F(k_f)$ factor defined by Eq. (5.2). This correction is only necessary when scans are performed with E_i fixed, so that k_f varies with ΔE. The most significant part of the correction is the factor $k_f^3 / \tan \theta_A$, which accounts for the dependence of the resolution volume on k_f. The rapid variation of this factor with ΔE can be rather inconvenient for measurements in energy-loss mode ($E_f < E_i$). A practical, although not absolute, guideline is that ΔE be no greater than 40% of the incoming energy. A demonstration of the resolution factor for energy-loss measurements is shown in Fig. 5.6. The resolution factor alone gives an adequate description of the k_f dependence of these normalized integrated intensities. For measurements over a larger range of k_f, it is necessary to correct for the gradual decrease in analyzer reflectivity with increasing k_f. Sharp dips can also occur over narrow k_f ranges due to competing Bragg reflections (Chesser and Axe, 1973).

For measurements which require an especially low background, the E_i-fixed mode is advantageous. Higher-order neutrons can be filtered out of the incident beam, so that they cannot contribute to the background through scattering. Of course, the k_f dependence is a disadvantage for energy-loss measurements over a wide range of ΔE, as we have just discussed. This problem is compounded for phonons (and magnons in antiferromagnets) because the cross section falls off with increasing ΔE due to the Bose factor and the $1/\hbar\omega$ factor, along with the resolution volume. Nevertheless, one can always compensate by adjusting the counting time, keeping in mind that longer counting times will also cause the net background level to increase. On the other hand, for measurements in energy-gain mode the variation in resolution volume will help to compensate for the dependence of the cross section on $\hbar\omega$. Energy-gain measurements are practical when $\Delta E < k_B T$.

5.1.2 E_f-fixed mode

This is the scattering mode more frequently used for phonon measurements. The factor $F(k_f)$ becomes a constant, since k_f is held fixed. This does not

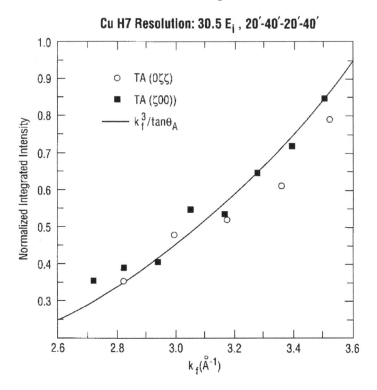

Fig. 5.6. Demonstration of k_f dependence of integrated intensities measured by energy loss with E_i fixed. Results shown are for transverse acoustic (TA) phonons in Cu at room temperature with $E_i = 30.5$ meV. The integrated intensities have been normalized to account for the Q and ω dependencies.

mean that the resolution volume is not changing; rather, the variation is compensated for by counting for a fixed monitor count. (The counting time then automatically increases as the resolution volume decreases.) The only necessary correction for the integrated intensities is to adjust for the monitor sensitivity to higher-order neutrons in the incident beam. This correction can be large for $E_i < 25$ meV, but it can also be accurately calibrated, as discussed in §4.9. It can either be applied to the data after the measurement, or it can be accounted for by the data-collection program through an adjustment of the monitor rate as a function of E_i. The lack of dependence of integrated intensities on E_i is demonstrated in Fig. 5.7. The results shown are for TA [110]T_2 phonons in Cu, with the higher-order-neutron correction applied, and the intensities normalized to account for the Q and $\hbar\omega$ dependencies.

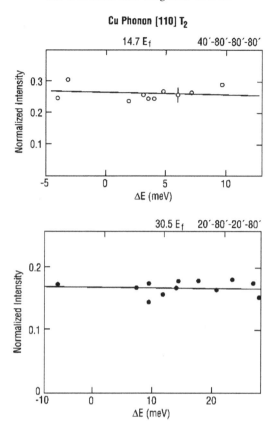

Fig. 5.7. Integrated intensities for Cu phonons measured in constant-**Q** mode with E_f fixed. The intensities have been corrected for the sensitivity of the monitor to higher-order neutrons in the incident beam, and they have been normalized to account for the Q and $\hbar\omega$ dependencies.

5.2 Phonons and magnons in iron

At room temperature Fe has the body-centered-cubic structure [see Fig. 5.8(a)] with

$$a = 2.866 \text{ Å}, \qquad \text{space group} = Im3m.$$

The extinction rule for Bragg reflections from a bcc lattice is

$$h + k + l = 2n.$$

The (hhk) zone [$(1\bar{1}0)$ plane] of reciprocal space is shown in Fig. 5.8(b). This choice of zone is a convenient one for phonon and magnon studies, because it allows simultaneous access to the [001], [110], and [111] symmetry

Fe

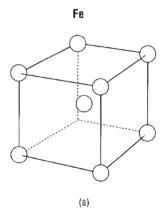

(a)

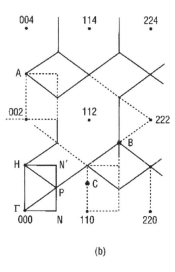

(b)

Fig. 5.8. (a) Unit cell of the bcc structure. (b) $(1\bar{1}0)$ plane of the reciprocal lattice for the bcc structure. High-symmetry points are labeled with letters.

directions. The Brillouin zone about each reciprocal-lattice point is indicated by a solid line. Note that the point (100) sits on the zone boundary.

Like Cu, Fe is a strong scatterer ($b = 0.95 \times 10^{-12}$ cm), making it very easy to measure phonons. The phonon dispersion curves obtained by Minkiewicz, Shirane, and Nathans (1967) are shown in Fig. 5.9. These results show all of the phonon branches that can be measured for a monatomic bcc lattice in the (hhk) zone. Note that along the [001] and [111] directions the two transverse modes are degenerate; however, along [110] they split apart. The two distinct modes are designated T_1 and T_2 in Fig. 5.9. The eigenvector of

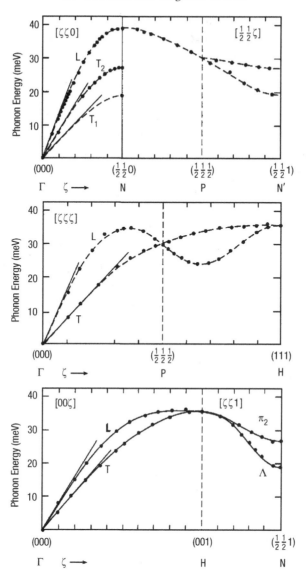

Fig. 5.9. Phonon dispersion curves in bcc Fe at room temperature as measured by Minkiewicz, Shirane, and Nathans (1967).

the T_1 mode is along [1$\bar{1}$0], which is perpendicular to the (1$\bar{1}$0) plane, and hence not observed. The other mode, T_2, has its eigenvector along [001], so that it can be measured in the (*hkk*) zone. As a result of the cubic symmetry, the initial slope of the [110]T_2 mode is equal to that of the [100]T mode.

Now let us consider the excitations (magnons) associated with the ferromagnetic order in Fe. With a Curie temperature of 768 °C (1041 K), the magnetic spin waves are quite well defined at room temperature. The integrated intensity for a constant-Q scan through a magnon dispersion curve in energy-loss mode is given by

$$\mathscr{I}_E \sim F(k_f)[n(\omega) + 1]S[gf(\mathbf{Q})]^2 e^{-2W(\mathbf{Q})}, \tag{5.6}$$

where $f(\mathbf{Q})$ is the magnetic form factor, and $\mu = gS\mu_B$ is the magnetic moment, with g being the Landé g-factor, S being the atomic spin, and μ_B representing a Bohr magneton. The other quantities are the same as in the formula for phonon intensity, Eq. (5.1). In principle, we should also include a factor to account for the $\mathbf{Q}$ dependence of magnetic scattering, i.e., only the components of the magnetic moments (or in this case the displacements of the magnetic moments) perpendicular to $\mathbf{Q}$ contribute to the scattered signal. However, for a cubic ferromagnet in zero field (i.e., with a random distribution of magnetic domains) this factor averages to a constant.

The magnetic moment of bcc Fe is reasonably large (2.2 μ_B), so that the magnons are easily detectable. However, the magnetic form factor decreases rapidly with increasing Q, which limits the range of reciprocal space that is useful for magnon measurements to the first two zone centers, (110) and (002). The spin-wave dispersion is described fairly well at small q by the simple dispersion relation

$$\hbar\omega_{\mathbf{q}} = Dq^2, \tag{5.7}$$

with

$$D \approx 280 \, \text{meV} \, \text{Å}^2.$$

As illustrated in Fig. 5.10(a), the magnon dispersion curve is rather steep, except at very small q values. For the higher-energy excitations, constant-Q scans give very broad peaks. It turns out to be more practical to perform scans in constant-E mode, which can yield sharp, well-defined peaks as shown in Fig. 5.10(b).

For determining dispersion curves, one simply needs to determine a peak position, and it is reasonable to choose between constant-Q and constant-E scans according to which mode gives the more precise peak definition. However, when it comes to evaluating integrated intensities, the two modes are no longer equivalent. The difference, as discussed in §4.7, is that the integrated intensity obtained from a constant-E scan depends inversely on the slope of the dispersion. This feature is demonstrated in Fig. 5.11. The top part of the figure shows the integrated intensities, corrected for the temperature factor $\langle n \rangle \equiv n(\omega) + 1$, for a series of constant-E scans. (Note that other than the temperature factor, there is no ω dependence in the

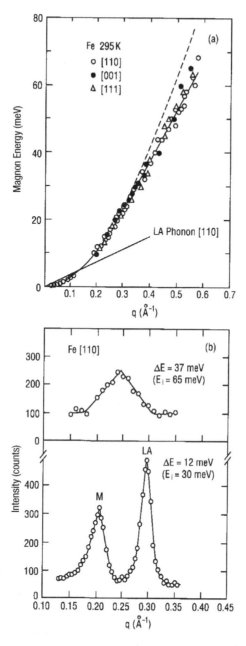

Fig. 5.10. (a) Dispersion relation for Fe at 295 K. The broken line corresponds to Eq. (5.7) with $D = 281\,\text{meV}\,\text{Å}^2$. (b) Constant-$E$ scans of spin waves (M) and longitudinal acoustic (LA) phonons in Fe at 295 K with fixed incident energy E_i (from Shirane, Minkiewicz, and Nathans, 1968).

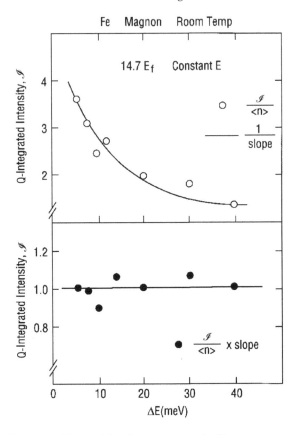

Fig. 5.11. (*Top*) integrated intensities for magnons in Fe at room temperature measured in constant-E mode. The intensities have been corrected for the Bose factor.
The solid line is proportional to the inverse of the slope of the energy dispersion.
(*Bottom*) Data from the top panel multiplied by the slope. $<n>$ is the temperature
factor (see text).

integrated intensity, and the variations in the magnetic form factor and the
Debye–Waller factor are negligible over the range of these measurements.)
Given the dispersion relation in Eq. (5.7), the slope is $2Dq = 2\sqrt{D\Delta E}$. As
one can see, the integrated intensity falls off as the inverse of the slope. The
lower part of the figure shows the result of correcting the intensities for the
slope dependence. The corrected intensities are now essentially independent
of energy, as expected.

We end this section with a brief discussion of the coupling between the
magnetic order and the elastic properties in Fe. Since the elastic constants
are related to second derivatives of the crystal's potential energy, which in
turn has a significant contribution associated with the magnetic ordering,

it should not be surprising that the elastic constants c_{ij} show anomalies at T_C. The largest effect is observed for $c' = (c_{11} - c_{12})/2$, which is the elastic constant corresponding to the small-q slope of the $[110]T_1$ phonon mode. Figure 5.12(a) compares the ultrasonic measurements of Dever (1972) with the neutron scattering results obtained by Satija, Comès, and Shirane (1985). As one can see, there is a very large temperature dependence of c', with an inflection point at T_C.

The surprising result obtained in the neutron study is that the strong softening with temperature of the $[110]T_1$ mode extends all the way to the zone boundary, as shown in Fig. 5.12(b). However, despite the strong softening, the phonon profile remains sharp, as illustrated in Fig. 5.13. No anomaly was observed in the phonon linewidth at T_C.

5.3 Magnons in MnF₂

MnF_2 is a classic material for the study of spin waves in an antiferromagnet. The crystal structure is tetragonal, with

$$a = 4.873 \text{ Å}, \qquad \text{space group} = P4_2/mnm,$$
$$c = 3.130 \text{ Å}.$$

The Mn ions form a body-centered-tetragonal lattice that is considerably compressed in the c-direction, while the F atoms occupy non-centrosymmetric positions between them, as indicated in Fig. 5.14. The atomic positions are:

Mn: $(2a)$ $0, 0, 0$; $\frac{1}{2}, \frac{1}{2}, \frac{1}{2}$.

F: $(4f)$ $x, x, 0$; $\bar{x}, \bar{x}, 0$; $\frac{1}{2} + x, \frac{1}{2} - x, \frac{1}{2}$; $\frac{1}{2} - x, \frac{1}{2} + x, \frac{1}{2}$.

The only constraint on Bragg reflections is

$$h0l: \quad h + l = 2n$$

The shortest Mn–Mn distance is along the $\langle 001 \rangle$ direction, with next-nearest neighbors along $\langle 111 \rangle$, and third neighbors along $\langle 100 \rangle$ and $\langle 010 \rangle$.

Below the Néel temperature of 67.5 K the Mn moments order in a simple antiferromagnetic structure, with all spins pointing along the c-axis, and the spin at the center of a unit cell pointing opposite to those in the corners (see Fig. 5.14). The extinction rule for the magnetic scattering peaks is

$$h + k + l = 2n + 1.$$

The magnetic reflections will tend to overlap with nuclear reflections, except for the case $k = 0$. Thus, it is convenient to study the magnetic scattering

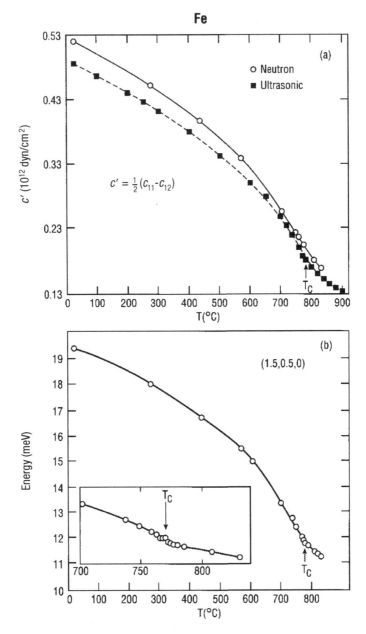

Fig. 5.12. (a) The elastic constant $c' = (c_{11} - c_{12})/2$ for Fe as a function of temperature. The ultrasonic measurements are from Dever (1972), and the neutron results are from Satija, Comès, and Shirane (1985). (b) Energy of the [110]T_1 mode at the zone boundary [measured at reciprocal point (1.5,0.5,0)] as a function of temperature. The inset shows detailed measurements of the zone-boundary frequency in the vicinity of T_C (from Satija, Comès, and Shirane, 1985).

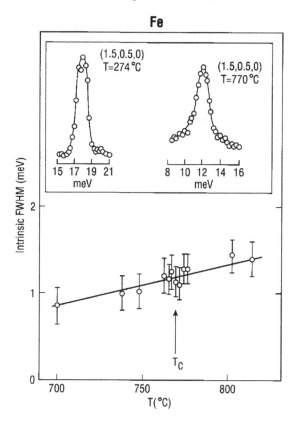

Fig. 5.13. Intrinsic full-width at half-maximum for the zone-boundary $[110]T_1$ phonon in Fe as a function of temperature. The inset shows the phonon profile at $T = 274$ and $770\,°C$ (near T_C) (from Satija, Comès, and Shirane, 1985).

in the $(h0l)$ zone, which is illustrated in Fig. 5.15. This zone is also useful because it allows one to probe spin waves propagating both parallel and perpendicular to the c-axis.

Turning to the cross section for exciting spin waves in MnF_2, the integrated intensity for a constant-$\mathbf{Q}$ scan may be written

$$\mathscr{I}_E \sim F(k_f)[n(\omega) + 1]S|gf(\mathbf{Q})|^2 \left(1 + \frac{Q_z^2}{Q^2}\right) t^2(\mathbf{Q})e^{-2W(\mathbf{Q})}. \qquad (5.8)$$

The factor $[1 + (Q_z/Q)^2]$ takes account of the fact that neutrons scatter only from the components of the magnetic moments that are perpendicular to $\mathbf{Q}$. The spin waves are transverse excitations, by definition, so the neutrons will scatter from the displacements of the spins away from their average orientation, which is along the c-axis. Thus, if $\mathbf{Q} \parallel \mathbf{c}$ one will observe spin waves with displacements along both $\mathbf{a}$ and $\mathbf{b}$, but if $\mathbf{Q} \parallel \mathbf{a}$ only the mode

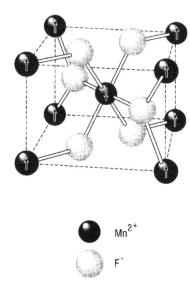

Mn²⁺

F⁻

Fig. 5.14. Crystal structure of MnF$_2$, with the magnetic structure indicated by the superimposed arrows.

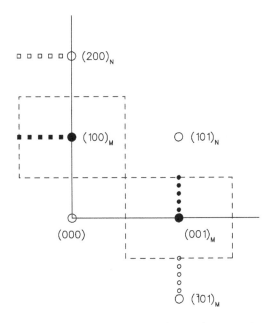

Fig. 5.15. Diagram of the (*h0l*) zone of reciprocal space for MnF$_2$. The large solid (open) circles indicate nuclear (antiferromagnetic) reflections. The dashed lines denote antiferromagnetic Brillouin zone boundaries. Small circles (squares) indicate positions for well-focused constant-**Q** scans of spin waves propagating in the [100] ([001]) direction.

with displacements along **b** will be seen. The major generic difference from the formula for ferromagnetic magnons is the factor $t^2(\mathbf{Q})$ which, for small q in an antiferromagnetic zone, is given by

$$t^2(\mathbf{Q}) \approx \frac{2\omega_2}{\omega_\mathbf{q}}, \qquad (5.9)$$

where $\hbar\omega_2$ is a characteristic energy to be defined below. Since the magnetic form factor $f(\mathbf{Q})$ restricts measurements to small Q, and the Néel temperature limits studies of the ordered state to low temperatures, the Debye–Waller factor can be approximated as unity for all practical purposes.

Before describing the spin-wave dispersion formula, we first consider the effective couplings responsible for it. The dominant magnetic interaction, which we denote by J_2, is superexchange between second-nearest-neighbor Mn moments via the fluorine ligands. It is interesting to note that while a superexchange interaction via a ligand positioned symmetrically between magnetic ions would lead to zero moment on the ligand, the asymmetrical distribution of Mn moments about a fluorine site results in a net spin density on the F sites, as shown by Nathans *et al.* (1963). Nearest-neighbor (along $\langle 001 \rangle$) Mn moments interact through a much weaker coupling J_1, which has been shown to be ferromagnetic by analysis of spin-wave measurements (Okazaki, Turberfield, and Stevenson, 1964). The same study showed that interactions between third neighbors make an insignificant contribution to the dispersion. However, there is a substantial single-ion anisotropy energy D, which can largely be explained by dipole–dipole interactions (Keffer, 1952). This description can be summarized by the following effective spin hamiltonian:

$$H = \tfrac{1}{2}J_2 \sum_{\mathbf{r},\mathbf{d}_2} \mathbf{S}_\mathbf{r} \cdot \mathbf{S}_{\mathbf{r}+\mathbf{d}_2} - \tfrac{1}{2}J_1 \sum_{\mathbf{r},\mathbf{d}_1} \mathbf{S}_\mathbf{r} \cdot \mathbf{S}_{\mathbf{r}+\mathbf{d}_1} - D \sum_\mathbf{r} (S_{\mathbf{r},z})^2, \qquad (5.10)$$

where $\mathbf{d}_n$ labels the nth nearest-neighbors of Mn site $\mathbf{r}$, and J_1 and J_2 are assumed to have positive values.

The formula for spin-wave dispersion corresponding to the effective spin hamiltonian is (Okazaki, Turberfield, and Stevenson, 1964)

$$\hbar\omega_\mathbf{q} = \hbar\omega_2 \left[(1 + \zeta_\mathbf{q})^2 - \gamma_\mathbf{q}^2 \right]^{\frac{1}{2}}, \qquad (5.11)$$

where

$$\zeta_\mathbf{q} = \left[D + 2\hbar\omega_1 \sin^2 \left(\tfrac{1}{2}q_z c \right) \right] / \hbar\omega_2, \qquad (5.12)$$

$$\gamma_\mathbf{q} = \cos \left(\tfrac{1}{2}q_x a \right) \cos \left(\tfrac{1}{2}q_y a \right) \cos \left(\tfrac{1}{2}q_z a \right), \qquad (5.13)$$

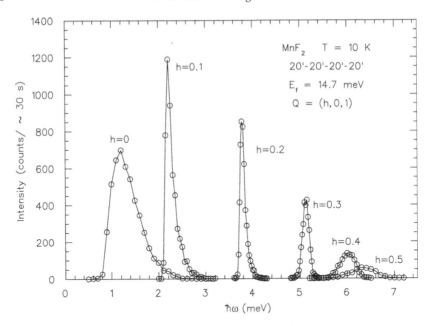

Fig. 5.16. Constant-**Q** scans of spin waves in a large single crystal of MnF_2 at 10 K. The measurement at each energy point was for a fixed monitor count, that was adjusted to correct for the monitor's sensitivity to harmonics in the incident beam.

and

$$\hbar\omega_i = 2Sz_iJ_i. \tag{5.14}$$

Here a and c are the lattice parameters, $z_1 = 2$ is the number of nearest neighbors, and $z_2 = 8$ is the number of second-nearest neighbors. If D and J_1 were zero, then $\zeta_{\mathbf{q}}$ would be zero, and for small q one would have

$$\frac{\omega_{\mathbf{q}}}{\omega_2} = \sqrt{1 - \gamma^2} \approx \frac{qa}{2\sqrt{2}}, \tag{5.15}$$

ignoring the tetragonal distortion of the lattice. The anisotropy energy D introduces a gap at $q = 0$, while a finite J_1 causes the dispersion to be different along the [100] and [001] directions.

In contrast to Fe, the energy scale for spin waves in MnF_2 is quite small, so that the dispersion can easily be mapped out across the entire Brillouin zone. Also, the scattering is reasonably strong because of the large Mn moment ($S = \frac{5}{2}$). One can best take advantage of the focusing effect by performing constant-**Q** scans with $\mathbf{q} \perp \mathbf{Q}$, as indicated in Fig. 5.15. As examples of this, Fig. 5.16 shows several scans of magnons collected near (001) with $\mathbf{q} \parallel [100]$, at **Q**s corresponding to the small filled circles in Fig. 5.15. The resulting

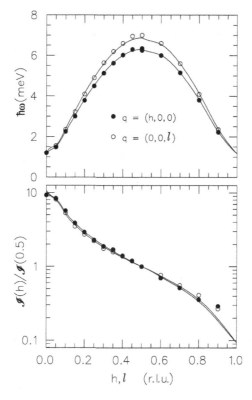

Fig. 5.17. *Top panel:* points indicate experimental results for spin-wave dispersion in MnF$_2$ along two directions in reciprocal space obtained from scans such as those shown in Fig. 5.16. The solid lines are fits using Eq. (5.11). *Bottom panel:* corresponding integrated intensities, corrected for the magnetic form factor and the Q_z-dependent factor in Eq. (5.8) and normalized at the zone-boundary points. The curves are calculated from Eq. (5.16). [r.l.u. = reciprocal lattice units.]

dispersion curve, along with results for $\mathbf{q} \parallel [001]$, is plotted in the top panel of Fig. 5.17. A very narrow magnon peak is observed where the dispersion is linear, but the focusing disappears at the zone center and zone boundary, where the dispersion flattens out. The peak profiles measured at the latter points are quite asymmetric due to the curvature of the dispersion.

To extract the most precise results from scans such as those in Fig. 5.16, one must fit the data by numerically convolving a model cross section with the resolution function. For the present purposes it is sufficient to determine $\hbar\omega_{\mathbf{q}}$ from the position of the intensity maximum in each scan. Fits to the experimental points using Eq. (5.11), indicated by the solid lines in the upper panel of Fig. 5.17, yield the results $J_1 = 0.032$ meV, $J_2 = 0.155$ meV, and $D = 0.11$ meV, which are consistent with the values obtained by Okazaki,

Turberfield, and Stevenson (1964). The figure illustrates the fact that the spin-wave dispersion is identical in antiferromagnetic and ferromagnetic Brillouin zones, although the intensities are not, as we discuss next.

From the integrated intensities, we can learn more about the dynamic structure factor, $t^2(\mathbf{Q})$. Looking back at Eq. (5.8), we see that besides $t^2(\mathbf{Q})$, other sources of $\mathbf{Q}$ dependence are the magnetic form factor and the factor $[1 + (Q_z/Q)^2]$. Correcting for the latter factors, and keeping in mind that, at 10 K, $n(\omega)$ is essentially zero for the energies of interest, we obtain the results shown in the bottom panel of Fig. 5.17, which should be proportional to $t^2(\mathbf{Q})$. These results can be compared with the following formula obtained from spin-wave theory:

$$t^2(\mathbf{Q}) = \frac{1 + \zeta_\mathbf{q} \pm \gamma_\mathbf{q}}{\sqrt{(1 + \zeta_\mathbf{q})^2 - \gamma_\mathbf{q}^2}}, \tag{5.16}$$

where the $+$ sign applies within an antiferromagnetic Brillouin zone and the $-$ sign in a ferromagnetic Brillouin zone. Curves calculated from this equation using the parameter values obtained from the dispersion measurements are shown in the figure. The experimental results are in good agreement with the theory, except near the ferromagnetic zone center, where the intensity is rather small and where resolution corrections may become more important. A more careful analysis has been made by Lurie *et al.* (1973), who used the extrapolation of the measured $t^2(\mathbf{Q})$ to a ferromagnetic zone center in order to obtain an absolute normalization of the dynamical spin susceptibility determined from the scattering intensities to the uniform susceptibility obtained from bulk measurements.

References

Chesser, N. J. and Axe, J. D. (1973). *Acta Cryst. A* **29**, 160.

Dever, D. J. (1972). *J. Appl. Phys.* **43**, 3293.

Keffer, F. (1952). *Phys. Rev.* **87**, 608.

Lurie, N. A., Shirane, G., Heller, P., and Linz, A. (1973). In *Magnetism and Magnetic Materials - 1972*, ed. C. D. Graham and J. J. Rhyne (American Institute of Physics, New York), p. 93.

Minkiewicz, V. J., Shirane, G., and Nathans, R. (1967). *Phys. Rev.* **162**, 528.

Nathans, R., Alperin, H. A., Pickart, S. J., and Brown, P. J. (1963). *J. Appl. Phys.* **34**, 1182.

Nicklow, R. M., Gilat, G., Smith, H. G., Raubenheimer, L. J., and Wilkinson, M. K. (1967). *Phys. Rev.* **164**, 922.

Okazaki, A., Turberfield, K. C., and Stevenson, R. W. H. (1964). *Phys. Lett.* **8**, 9.

Satija, S. K., Comès, R. P., and Shirane, G. (1985). *Phys. Rev. B* **32**, 3309.

Shirane, G., Minkiewicz, V. J., and Nathans, R. (1968). *J. Appl. Phys.* **39**, 383.

Svensson, E. C., Brockhouse, B. N., and Rowe, J. M. (1967). *Phys. Rev.* **155**, 619.

6

Spurious peaks

Because of the non-ideal behavior of various spectrometer components and the sample environment, scattering artifacts are often observed in both elastic and inelastic spectra. Such artifacts frequently appear as sharp, well defined peaks that mimic intrinsic scattering features. The proper identification of spurious peaks (sometimes facetiously referred to as "spurions") can be crucial to the successful interpretation of experimental measurements. In this chapter we will describe some of the possible sources of spurious scattering, and explain how to recognize and avoid such features. Once the cause of a particular artifact is understood, it may seem quite straightforward, even obvious; however, the initial discovery of the underlying mechanism has typically required clever detective work. New spurious features have been discovered as experiments have been pushed to increasingly higher levels of sensitivity. The information in this chapter has been slowly accumulated over the last several decades, and represents contributions by many different workers.

6.1 Higher-order neutrons

An ideal neutron monochromator would select a unique incident wave vector k_i. In real life one must make use of Bragg diffraction by a crystal, as discussed in Chap. 3. If the scattering planes of the monochromator crystal correspond to reciprocal-lattice vector G_M and the scattering angle is θ_M, then Bragg's law gives

$$k_i(n_M) = n_M G_M / 2 \sin \theta_M, \tag{6.1}$$

where, in general, n_M can be any positive integer. The spectral distribution of the source and the reflectivity of the monochromator determine the flux at each harmonic. While the flux of such higher-order neutrons with $n_M = 2, 3, \ldots$ is typically (though not always) hundreds of times weaker than that at the fundamental wave vector, they can nevertheless cause spurious peaks of comparable magnitude to weak intrinsic features. Below we discuss separately the cases of higher-order contributions in elastic and inelastic scattering.

6.1.1 Elastic scattering

Higher-order neutrons create spurious diffraction peaks when the monochromator and analyzer are the same type of crystal set for the identical reflection. In such a case, one will observe a sum of diffraction spectra corresponding to the fundamental wave vector plus the higher harmonics. When the diffraction pattern is interpreted in terms of the fundamental wave vector alone, then weak spurious superlattice peaks will occur at positions G/n, where G represents the wave vectors of the fundamental reflections. Such artifacts can lead to great confusion in studies of antiferromagnetic or charge-density-wave ordering, where weak superlattice peaks are anticipated.

To avoid any confusion, the higher-order neutrons should be eliminated from the incident beam. This is typically accomplished by using an appropriate filter, such as pyrolytic graphite or beryllium, as discussed in Chap. 3. Of course, a filter attenuates, but does not completely eliminate, the harmonics, and the effectiveness of a filter can vary with neutron energy. When intrinsic superlattice peaks are of interest, it is good practice to measure the degree of harmonic attenuation, so that the level of contamination can be anticipated. Alternatively, one can eliminate the second-order contribution by utilizing an odd reflection of a non-centrosymmetric monochromator crystal, such as Ge (111).

6.1.2 Inelastic scattering

For inelastic scattering, we must consider the possibility that incidental elastic scattering at the sample may reach the detector due to diffraction by different harmonics at the monochromator and analyzer. The n_Mth harmonic transmitted by the monochromator has an energy

$$E_i(n_\mathrm{M}) = n_\mathrm{M}^2 E_i, \tag{6.2}$$

and similarly for the analyzer,

$$E_f(n_\mathrm{A}) = n_\mathrm{A}^2 E_f. \tag{6.3}$$

Elastic scattering at the sample will contribute to the detected signal at an energy such that

$$E_i(n_\mathrm{M}) = E_f(n_\mathrm{A}), \tag{6.4}$$

Table 6.1. *Table of nominal excitation energies $\hbar\omega$, normalized to E_f, at which a false peak can occur due to higher-order diffraction at the monochromator and/or the analyzer. Values are given for various combinations of monochromator diffraction order n_M and analyzer order n_A. Bold values indicate conditions where a filter after the sample (standard E_f-fixed configuration) would not eliminate the spurious feature.*

n_M $n_A =$	1	2	3	4
1	0	3	8	15
2	$-\frac{3}{4}$	0	$\frac{5}{4}$	3
3	$-\frac{8}{9}$	$-\frac{5}{9}$	0	$\frac{7}{9}$
4	$-\frac{15}{16}$	$-\frac{3}{4}$	$-\frac{7}{16}$	0

even though $E_i \neq E_f$. For identical monochromator and analyzer crystals set to the same reflection, a spurious signal will occur when the spectrometer is set for an energy transfer

$$\hbar\omega = E_i - E_f \tag{6.5}$$

$$= \left(1 - \frac{n_M^2}{n_A^2}\right) E_i \tag{6.6}$$

$$= \left(\frac{n_A^2}{n_M^2} - 1\right) E_f, \tag{6.7}$$

where we have made use of Eqs. (6.2)–(6.4).

For example, suppose one is working with a fixed initial energy of 40 meV. When the analyzer is set to detect 10 meV neutrons, it will also transmit neutrons at 40 meV by second-order diffraction. Thus, any elastic incoherent scattering at the sample will result in a spurious signal at the nominal energy transfer of 30 meV. For such a fixed-E_i measurement, the standard use of a filter in the incident beam would have no effect on this process. When both n_M and $n_A \geq 2$, a filter will attenuate the elastic scattering process; nevertheless, an incidental Bragg reflection at the sample could result in a significant peak. Such a "spurion" will be sharp in energy, and can only occur at specific energy transfers. The most significant false-peak conditions are tabulated in Table 6.1, for E_f-fixed mode, and Table 6.2, for E_i-fixed mode. The information in Table 6.1 is depicted graphically in Fig. 6.1.

The false-peak problem can be avoided by following two basic rules: (1) The neutron energy (both E_i and E_f) should be as low as possible, in order to minimize the possibility of incidental Bragg scattering at the sample.

Table 6.2. *Same as Table 6.1, but with $\hbar\omega$ normalized to E_i.*

n_M $n_A =$	1	2	3	4
1	0	$\frac{3}{4}$	$\frac{8}{9}$	$\frac{15}{16}$
2	-3	0	$\frac{5}{9}$	$\frac{3}{4}$
3	-8	$-\frac{5}{4}$	0	$\frac{7}{16}$
4	-15	-3	$-\frac{7}{9}$	0

(2) The nominal energy transfer $\hbar\omega$ should be no larger than roughly half of E_i. These rules tend to conflict with each other, and cannot always be satisfied conveniently. When they are violated, the false-peak conditions should be kept in mind.

A more subtle contribution due to harmonics in the incident beam can occur when excitations exist in the sample at an energy much greater than the nominal energy transfer. An example of such a spurion was observed in a study of lattice dynamics in $NbD_{0.85}$ by Shapiro *et al.* (1978). In measurements with $E_f \approx 15\,\mathrm{meV}$, a dispersionless peak was observed at an energy transfer of 10 meV. It was later realized (Shapiro *et al.*, 1981) that the apparent 10-meV peak actually corresponded to a local mode near 85 meV that was observed via second-order neutrons in the incident beam. To see this, note that with $\hbar\omega = 10\,\mathrm{meV}$ and $E_f = 15\,\mathrm{meV}$, the incident energy would be set to $E_i = 25\,\mathrm{meV}$. The energy transfer corresponding to the second-order neutrons is then

$$\begin{aligned}
\hbar\omega' &= 4E_i - E_f \\
&= 100 - 15 \\
&= 85\,\mathrm{meV}.
\end{aligned}$$

To demonstrate that such a feature is spurious, one can vary E_f, or use fixed-E_i mode with a filter before the sample.

6.2 Accidental Bragg scattering

A triple-axis spectrometer is designed to measure inelastic scattering from a sample crystal utilizing Bragg scattering from the monochromator and analyzer crystals. However, the intended scattering event is only one of several more-general possibilities. Whenever two out of three of the crystals are set to Bragg reflect, a weak scattered signal from the third crystal may be detected. In particular, if the sample angle accidentally corresponds to

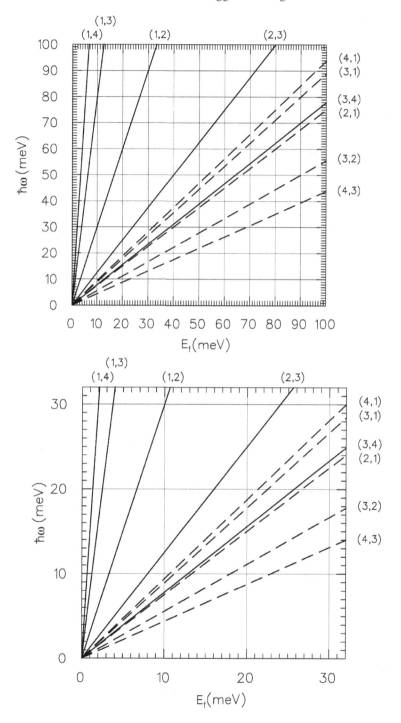

Fig. 6.1. Graphical presentation of the false-peak conditions listed in Table 6.1. The diffraction orders for each condition are denoted as (n_M, n_A).

a Bragg reflection, then incoherent or thermal diffuse scattering from either the monochromator or analyzer may result in a spurious peak. While this situation seems obvious in hindsight, a long and persistent effort was required to properly identify such artifacts. At Brookhaven, the problem of accidental Bragg scattering from the sample was first documented and explained by Currat and Axe (1978), and, as a result, such features are frequently referred to as Currat–Axe peaks.

In the general case of accidental Bragg scattering, spurious peaks disperse along particular lines in reciprocal space. In many cases, these "false" dispersion curves will occur along directions unrelated to the crystal symmetry, and it is fairly unusual for these "extra" dispersion branches to be observed. A more common problem is for a single spurious peak to appear due to accidental Bragg scattering. Such peaks are most dangerous when they occur in a region where one hopes to find structure. The problem of spurious dispersion curves becomes more serious when the ratio of q to the length of the reciprocal-lattice vector G becomes comparable to the mosaic width of the sample. The observation of unexpected, linearly-dispersing peaks in a study of spin-density waves in Cr motivated investigations by Uemura, Grier, and Shirane (1982) and by Currat and Dorner (1982). A detailed discussion of the problem is given in the latter work.

6.2.1 General case

Consider the scattering diagram shown in Fig. 6.2(a). (For historical reasons, we will work with a left-handed coordinate system (see Fig. 4.2(d)–(f)) throughout this section.) The spectrometer is set to measure energy-loss scattering at a momentum transfer $\mathbf{Q}$. However, the geometry is such that elastic scattering by the sample, corresponding to the reciprocal-lattice vector $\mathbf{G}$, will be in the same direction as the nominal $\mathbf{k}_f$. Thus, if there is some incoherent or diffuse scattering by the analyzer crystal, some of the neutrons scattered elastically by the sample will reach the detector along with the intended signal. The accidental scattering is not limited to energy-loss mode; if we change the length of k_f but keep the angles and k_i fixed, the accidental Bragg scattering will remain. The situation for energy gain is shown in Fig. 6.2(d). Since this process involves weak scattering by the analyzer, we will label it a type-A event.

Because of the symmetry of a triple-axis spectrometer, it must be possible to have accidental Bragg scattering caused by weak scattering at the monochromator instead of the analyzer. Figure 6.2(b) shows such a scattering diagram for energy loss. Incoherent scattering at the monochromator

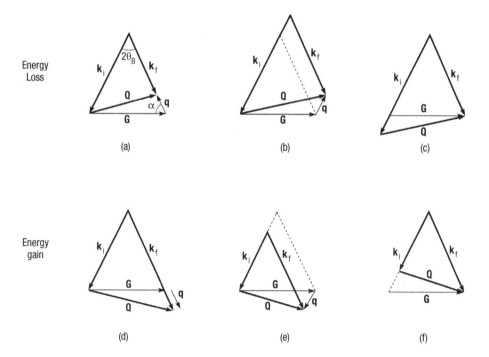

Fig. 6.2. Scattering diagrams illustrating the conditions for accidental Bragg scattering. Type-A processes are illustrated for energy-loss and energy-gain conditions in (a) and (d), respectively. Type-M processes are illustrated in (b) and (e). Figures (c) and (f) are the same as (b) and (e), except that the reciprocal-lattice vector **G** has been translated to emphasize the symmetry with (d) and (a), respectively.

can provide neutrons traveling parallel to $\mathbf{k}_i$, but with magnitude k_f, which will Bragg scatter from the sample with the same $\mathbf{k}_f$ as those due to the desired inelastic process. We will label this form of accidental Bragg scattering a type-M event. To emphasize the symmetry with the type-A process, we have redrawn the diagram in Fig. 6.2(c) with the reciprocal-lattice vector translated; note the symmetry with diagram (d). Similarly, diagram (f) is symmetric with (a).

To understand the significance of this spurious scattering, it is helpful to determine the points in $(\mathbf{Q}, \omega)$ space at which it may occur. Let us start with the type-A process. One condition for the accidental scattering is that the scattering angle θ_S be equal to the Bragg angle θ_B defined by $\sin \theta_B = G/2k_i$. The nominal incident and final neutron wavevectors are related to the reciprocal-lattice vector by

$$\mathbf{G} = \left(\frac{k_i}{k_f} \right) \mathbf{k}_f - \mathbf{k}_i. \tag{6.8}$$

As usual, we use the definitions $\mathbf{Q} = \mathbf{k}_f - \mathbf{k}_i$ and $\mathbf{q} = \mathbf{Q} - \mathbf{G}$. If we denote the angle between $\mathbf{q}$ and $-\mathbf{G}$ as α and allow q to take on negative magnitudes, then one can easily show that

$$q = k_i - k_f, \tag{6.9}$$

$$\cos\alpha = \sin\theta_B = G/2k_i. \tag{6.10}$$

From the definition $\hbar\omega = E_i - E_f$, where E_i and E_f are the nominal incident and final neutron energies, one finds that the type-A spurions have the effective dispersion relation

$$\hbar\omega_A = -\frac{\hbar^2}{m_n}k_iq\left(1 + \frac{q}{2k_i}\right). \tag{6.11}$$

For $q \ll k_i$ and fixed k_i the dispersion is linear in q, with a slope of $-4k_i\,\mathrm{meV\,\mathring{A}}$.

The analysis of the type-M process is quite similar. In this case θ_B is determined by k_f rather than k_i, and Eq. (6.8) becomes

$$\mathbf{G} = \mathbf{k}_f - \left(\frac{k_f}{k_i}\right)\mathbf{k}_i. \tag{6.12}$$

One finds

$$q = k_i - k_f, \tag{6.13}$$

$$\cos\alpha = -\sin\theta_B = -G/2k_f. \tag{6.14}$$

The effective dispersion is given by

$$\hbar\omega_M = -\frac{\hbar^2}{m_n}k_fq\left(1 - \frac{q}{2k_f}\right). \tag{6.15}$$

As indicated by Eqs. (6.10) and (6.14), the accidental scattering occurs along specific curves in reciprocal space whose shapes depend on whether one is working with E_i or E_f fixed. The two cases are illustrated in Fig. 6.3. With k_i constant, as in Fig. 6.3(a), the angle between $\mathbf{q}$ and $\mathbf{G}$ is a constant for type-A peaks, but it varies with k_f for type-M peaks. The situation is reversed for k_f fixed, as shown in Fig. 6.3(b). For small q the angle between the two curves is approximately $2\theta_B$, as indicated in the figure.

For inelastic measurements along symmetry directions with q not too small, the lines of accidental Bragg scattering will typically be rotated away from the region of study. Only under particularly unfortunate conditions will they be a problem. One such situation is illustrated in Fig. 6.4. Suppose one wishes to perform measurements near the (110) Bragg peak of a cubic crystal in E_f-fixed mode to study acoustic phonons propagating in the [100]

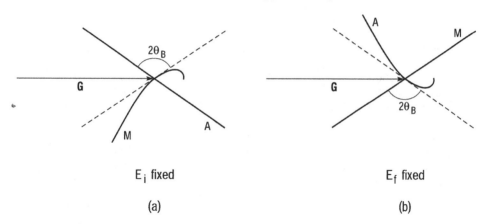

Fig. 6.3. Schematic diagrams of the lines in reciprocal space along which type-A and type-M accidental Bragg scattering can occur relative to a reciprocal-lattice vector **G**. (a) E_i-fixed mode. (b) E_f-fixed mode.

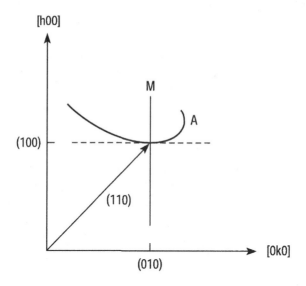

Fig. 6.4. Example of a situation in which the lines of accidental Bragg scattering could overlap with symmetry directions in a sample.

or [010] directions. If a is the lattice parameter and, by chance, $k_f = 2\pi/a$, then $\theta_B = 45°$, and the type-A and type-M curves will overlap the directions of interest.

To avoid, or be prepared for, spurious peaks, one can use a simple computer program to check for accidental Bragg scattering at each point in a constant-**Q** or constant-E scan. Such a program has been in use for

h	k	e	h1	k1	h2	k2
2.200	2.000	2.000	2.198	2.071	2.129	2.006
2.200	2.000	3.000	2.195	2.106	2.094	2.009
2.200	2.000	4.000	2.191	2.140	2.060	2.012
2.200	2.000	5.000	2.186	2.174	2.027	2.015
2.200	2.000	6.000	2.181	2.206	1.993	2.017
2.200	2.000	7.000	2.174	2.239	1.961	2.019
2.200	2.000	8.000	2.166	2.271	1.928	2.021
2.200	2.000	9.000	2.158	2.302	1.896	2.023
2.200	2.000	10.000	2.149	2.332	1.865	2.024
2.200	2.000	11.000	2.139	2.362	1.833	2.025
2.200	2.000	12.000	2.128	2.392	1.803	2.026
2.200	2.000	13.000	2.116	2.421	1.772	2.027
2.200	2.000	14.000	2.104	2.449	1.742	2.027
2.200	2.000	15.000	2.091	2.476	1.712	2.028
2.200	2.000	16.000	2.077	2.504	1.683	2.028
2.200	2.000	17.000	2.063	2.530	1.653	2.027
2.200	2.000	18.000	2.048	2.556	1.624	2.027
2.200	2.000	19.000	2.033	2.581	1.596	2.026
2.200	2.000	20.000	2.017	2.606	1.567	2.025

Fig. 6.5. Sample computer output from routine that checks for possible accidental Bragg scattering. The first three columns give the nominal components of $\mathbf{Q}$ (h and k, in reciprocal-lattice units) and energy (e in meV) for a scan. The columns labeled h1 and k1 give the values of h and k for a type-A scattering, while h2 and k2 correspond to a type-M scattering.

some time at Chalk River, and one was also implemented at Brookhaven. An example of the output is shown in Fig. 6.5. This example was calculated for a constant-$\mathbf{Q}$ scan of Cu ($a^* = 1.738\,\text{Å}^{-1}$) measured with k_f held fixed at $3.837\,\text{Å}^{-1}$ ($E_f = 30.5\,\text{meV}$). From an inspection of the columns labeled h2 and k2 one can see that conditions for a type-M event are very nearly satisfied when the energy transfer is $\sim 5.5\,\text{meV}$. Of course, because of finite resolution effects, it is not necessary to satisfy the conditions exactly in order to detect a spurious peak.

Currat and Axe (1978) suggested a particularly simple procedure for identifying spurious peaks *a posteriori*. The basic idea is that for accidental Bragg scattering occurring at nominally inelastic conditions $E_i^* \neq E_f^*$, the angles corresponding to spectrometer motors $C2$ and $A2$ must be the same as for Bragg scattering with either $E_i = E_f = E_i^*$ (type-A peak) or $E_i = E_f = E_f^*$ (type-M peak). The procedure is as follows:

(i) Note the values of the spectrometer parameters $(E_i^*, E_f^*, C2^*, A2^*)$ corresponding to the inelastic peak in question.

(ii) With the help of a rough scattering diagram, determine the reciprocal-lattice vector **G** believed to be involved.

(iii) Set the spectrometer for that particular Bragg reflection using first $E_i = E_f = E_i^*$ and then $E_i = E_f = E_f^*$, and record the values of $C2_i, A2_i$ and $C2_f, A2_f$, respectively.

(iv) Then $C2_i \approx C2^*$ and $A2_i \approx A2^*$ indicates the peak is of type-A, while $C2_f \approx C2^*$ and $A2_f \approx A2^*$ indicates type-M.

It was noted by Ishikawa, Fincher, and Shirane (1980) that type-A peaks can be a more serious problem than one might expect from the above discussion. Once the sample is oriented (accidentally) for Bragg scattering corresponding to a given k_i, spurious peaks can be observed even when $\mathbf{k}_i$ is offset in angle by as much as a few degrees from the "optimum" position. Because of the strength of the Bragg scattering relative to typical inelastic processes of interest, small-angle scattering from spectrometer components and imperfect collimation can conspire to deflect spurious neutrons into the detector. They found that spurious peaks due to accidental Bragg scattering can be reduced by tightening collimation immediately before and after the sample. Contributions due to small-angle scattering by a pyrolytic graphite filter in the scattered beam can be minimized by positioning the filter between the analyzer and detector.

6.2.2 Small q

The accidental Bragg scattering just discussed occurs only along specific curves in reciprocal space, and, in particular, it will not occur in the longitudinal direction. Nevertheless, it was known for quite a long time that spurious dispersion curves can be observed in the longitudinal direction in high-resolution measurements close to a Bragg peak. An example is given in Fig. 6.6 which shows the positions of peaks measured in constant-E scans for a Cr single crystal at room temperature (Uemura, Grier, and Shirane, 1982). Chromium metal is an incommensurately-ordered antiferromagnet. The measurements shown were made near the magnetic peak at $\mathbf{Q} \approx (0.96, 0, 0)$ in reciprocal-lattice units, with $a^* = 2.178$ Å^{-1}. The spin waves disperse very steeply, as indicated by the dashed line, and are not easily resolved. The peaks dispersing with a finite slope are the spurious features. Note that the slope changes as the energy of the scattered neutrons is changed – this is a clear sign that the observed features are not intrinsic to the sample.

The analysis of Currat and Dorner (1982) makes the explanation of the spurious peaks quite obvious. Any real crystal will have a finite mosaic

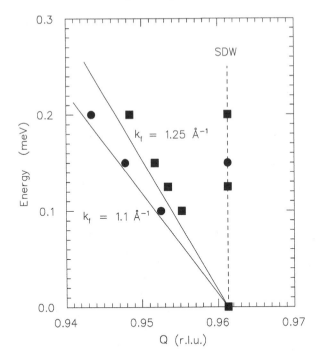

Fig. 6.6. Positions of peaks measured in constant-E scans near a $(Q_0, 0, 0)$ incommensurate magnetic peak corresponding to spin-wave-density (SDW) order in Cr metal at room temperature. The measurements were performed by Uemura, Grier, and Shirane (1982) at the H9 spectrometer at the HFBR. [r.l.u. = reciprocal-lattice units.]

distribution whose angular FWHM we will label η. The problem occurs when measurements are made at values of q such that $q/G \lesssim \eta/2$. For a mosaic crystal, the Bragg point spreads into an arc. As a result, as long as q is small enough, accidental Bragg scattering can occur for every $\mathbf{q}$, not just for $\mathbf{q}$ in special directions. The situation is illustrated in Fig. 6.7 for type-A scattering.

To calculate the energy at which the accidental type-A scattering will occur for a given $\mathbf{q}$, we first note that for a mosaic crystal Eq. (6.8) becomes

$$G = \left| \left(\frac{k_i}{k_f} \right) \mathbf{k}_f - \mathbf{k}_i \right|. \tag{6.16}$$

Combining this with the equation

$$\mathbf{Q} = \mathbf{G} + \mathbf{q} = \mathbf{k}_f - \mathbf{k}_i, \tag{6.17}$$

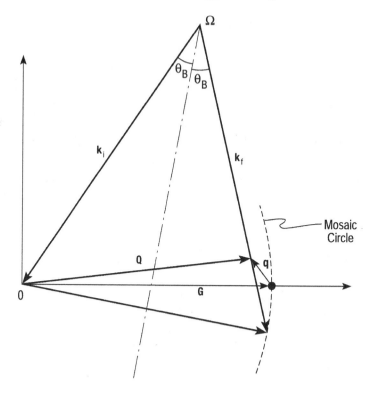

Fig. 6.7. Scattering diagram for accidental type-A Bragg scattering when q/G is comparable to the mosaic width η.

we can solve for k_f, and then determine the energy from

$$\hbar\omega = \frac{\hbar^2}{2m_{\mathrm{n}}}\left(k_i^2 - k_f^2\right).\tag{6.18}$$

To simplify the result, we keep terms of no higher than first order in the small parameters $(k_i - k_f)/k_i$ and q/G. The result is

$$\hbar\omega_{\mathrm{A}} \approx -4E_i\left(\frac{\mathbf{q}\cdot\mathbf{G}}{G^2}\right).\tag{6.19}$$

Peforming a similar calculation for type M modes yields

$$\hbar\omega_{\mathrm{M}} \approx 4E_f\left(\frac{\mathbf{q}\cdot\mathbf{G}}{G^2}\right).\tag{6.20}$$

Note that to the level of approximation used here, E_i and E_f are interchangeable in these two formulas.

The anomalous dispersion curves shown in Fig. 6.6 correspond to type-A scattering. From Eq. (6.19) the dispersion should be

$$\hbar\omega_A \approx 4E_f \frac{q}{G}. \tag{6.21}$$

The solid lines in the figure represent the calculated type-A dispersion, which is in good agreement with the measurements. Type-M scattering should yield dispersion branches at positive $\mathbf{q} \cdot \mathbf{G}$, but none were observed in the measurements at the H9 spectrometer at the HFBR. The reason is that H9 had a two-crystal (double) monochromator. The constraint that a neutron must scatter from both monochromator crystals in order to reach the sample effectively eliminates any non-Bragg scattering contributions in the monochromator that might result in accidental Bragg scattering at the sample.

According to Eqs. (6.19) and (6.20), the energy of the spurious scattering depends only on the projection of $\mathbf{q}$ onto $\mathbf{G}$. Currat and Dorner (1982) verified this behavior through measurements near the same $(0.96, 0, 0)$ incommensurate magnetic peak in Cr studied by Uemura, Grier, and Shirane (1982). They performed a series of constant-E measurements in the $(h, 0, l)$ zone in the region of the peak, and verified that the accidental scattering occurs along lines parallel to $(0, 0, l)$ at the predicted values of h.

6.3 Elastic streaks at small q

The pursuit of weak signals can often lead to the discovery of new artifacts. An example of this is the search for charge-density-wave scattering in bcc potassium (Werner, Giebultowicz, and Overhauser, 1987). Peaks due to charge-density-wave ordering are predicted to occur very close to the (110) Bragg peak, with an intensity $\sim 10^{-5}$ that of the (110) reflection. Controversy over the interpretation of the experimental measurements led Werner and Arif (1988) to analyze the possibility of anisotropic elastic scattering in the neighborhood of a Bragg peak. They concluded that, at a sufficiently high level of sensitivity, one should observe a pair of streaks intersecting at the Bragg point and positioned symmetrically about the reciprocal-lattice vector $\mathbf{G}$, as illustrated in Fig. 6.8. The angle χ between the two streaks should typically be $\sim 2\theta_B$. The streaks are attributed to neutrons that manage to penetrate the blades of the Soller collimators or that undergo small-angle scattering in the sample environment plus Bragg diffraction at the sample. Thus, they are essentially a resolution effect associated with non-gaussian contributions to the transmission functions of the spectrometer components. For further details, the reader is referred to the article by Werner and

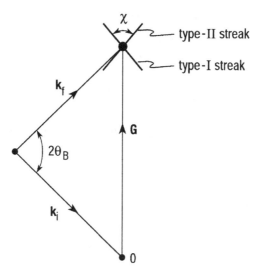

Fig. 6.8. Orientation in reciprocal space of elastic streaks due to non-ideal collimator performance and small-angle scattering.

Arif (1988). The contributions due to small-angle scattering have also been discussed by Pintschovius *et al.* (1987).

6.4 Resolution function artifacts

Artifacts associated with the resolution function should not be considered spurious, since they involve the intrinsic cross section of the sample and the non-isotropic shape of the resolution function. Nevertheless, the occurrence of Bragg-tail peaks and "forbidden" modes in inelastic scattering can cause significant consternation the first time one confronts them. Although we have already discussed these features in the previous chapter, we review them here for the sake of completeness.

6.4.1 Bragg tail

When performing an inelastic scan close to a Bragg reflection, a Bragg-tail peak will be observed if the tail of the resolution function sweeps through the corresponding reciprocal-lattice point. Because of the shape and typical orientation of the resolution function, the Bragg-tail scattering appears to disperse like a transverse acoustic phonon. The intensity of the scattering drops off rapidly as **q** and ω are increased. It is good practice to locate and quickly map out the Bragg-tail signal before starting inelastic measurements near a Bragg peak.

6.4.2 *"Forbidden" modes*

A common problem is to observe some signal from transverse phonons in a scan intended to probe only longitudinal modes. The scattering cross section for phonons is proportional to the square of the projection of the phonon eigenvector onto $\mathbf{Q}$. Along a high-symmetry direction in a crystal, one can choose $\mathbf{Q}$ to be parallel to longitudinal atomic displacements, so that transverse modes should be "forbidden." Of course, the relaxed vertical resolution causes a range of $\mathbf{Q}$s with significant transverse components to be sampled. This problem is particularly acute in measurements of longitudinal acoustic phonons close to a Bragg peak. The important point is to be aware of this phenomenon, so as to avoid misidentification of features.

We, of course, are explaining this problem with the benefit of hindsight. Originating the proper explanation can be difficult even for expert scatterers. In an early study of Pb, Brockhouse *et al.* (1962) observed forbidden transverse modes in scans of longitudinal phonons. They incorrectly interpreted the forbidden signal as being due to a complicated double-scattering process involving elastic scattering from some reciprocal-lattice point combined with inelastic scattering from a transverse excitation. This explanation is still occassionally given. The proper explanation in terms of the resolution function was convincingly demonstrated by Skalyo and Lurie (1973) (see § 4.8).

6.5 Artifacts due to sample environment

A number of scattering artifacts can arise from the sample surroundings. The most obvious is Bragg scattering from the sample holder. More subtle background effects come from scattering by gas molecules near the sample. While such contributions are often at an insigificant level, they can become important when one is searching for an extremely weak cross section. Cooling through the gas condensation temperature can cause a surprisingly large change in the background that can easily be misinterpreted. In this section we will discuss contributions from several materials commonly found in the beam path.

6.5.1 *Al*

Because it is inexpensive and easily machinable, aluminum is commonly used for sample holders, sample cans, and cryostat radiation shields. Since it is crystalline, an Al sample holder will contribute scattering at particular Bragg angles; however, the face-centered-cubic structure of Al leads to relatively few diffraction lines. For powder diffraction studies, some researchers prefer

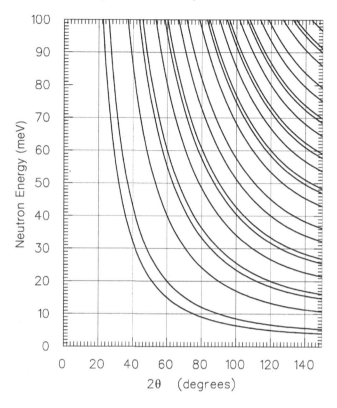

Fig. 6.9. Scattering angles of room-temperature aluminum Bragg peaks vs. neutron energy.

to use a sample container made of vanadium, which, by virtue of vanadium's negligible coherent cross section, eliminates the unwanted diffraction peaks. The downside for inelastic scattering is the larger incoherent cross section which results in an undesirable background.

In accepting the limitations of Al, it is good to keep in mind where the Bragg peaks occur. The scattering angles of the Bragg peaks at different neutron energies are plotted in Fig. 6.9. The Bragg angles for several popular neutron energies are listed in Table 6.3.

One word of caution on the use of Al is warranted. Aluminum stock metal is usually an alloy rather than the pure element. Aluminum alloys can show additional weak powder lines with intensities at the level of 10^{-3} of Al (111). The added elements also contribute a finite, though very weak, incoherent background. When a very low background is required, one should machine the sample holder from a pure-Al single crystal. Another characteristic of common Al alloys is the tendency to develop a preferred crystallographic

Table 6.3. *Scattering angles of Al Bragg reflections at particular neutron energies, using a = 4.0496 Å at 298 K.*

E	k	2θ$_B$ (deg)		
(meV)	(Å^{-1})	(111)	(200)	(220)
5.0	1.552	119.9	176.4	
13.7	2.570	63.1	74.3	117.3
14.7	2.662	60.6	71.3	111.1
30.5	3.834	41.0	47.7	69.8
41.0	4.445	35.2	40.9	59.2

orientation. Frequently one observes a strong Al (200) peak, but a very weak (111) reflection. This last effect causes no practical problem, but knowledge of the possibility of preferred orientation can prevent confusion.

6.5.2 Air

It is commonly recognized that air scattering can create a significant background signal. The majority of the signal comes from the region near the sample, and it can be sharply reduced by mounting the sample in an evacuated cryostat. To allow for good thermal conduction between the sample and cold finger in low-temperature experiments, it is standard practice also to enclose the sample in an Al can filled with He gas. Good conduction requires only that a large fraction of the exchange gas be He, and loading the gas in an unsealed glove-box frequently leaves 10 or 20% air in the sample can. At Brookhaven, we learned from hard experience that the residual air can lead to a very weak but temperature-dependent signature which can cause considerable confusion.

To characterize the air-scattering effect, we filled a standard sample can with one atmosphere of air at room temperature. The can was loaded into a cryostat, and the scattering was measured as a function of temperature using fairly open collimation (40'-40'-80'-80') and a fixed final energy of 14.7 meV. The observed scattering at $Q = 1.5$ Å^{-1} for energy transfers of 0, 3, and 6 meV is shown in Fig. 6.10. The scattering at 0 and 3 meV undergoes a sharp drop as one cools through 65 K. That drop corresponds to the condensation of N_2, which forms 78% of air by volume. The nitrogen scatters while in the gas phase, but is removed from the beam when it condenses onto the cold surface of the sample can. The condensation begins well below the familiar temperature of 77 K because the partial pressure of N_2 in the sealed can is

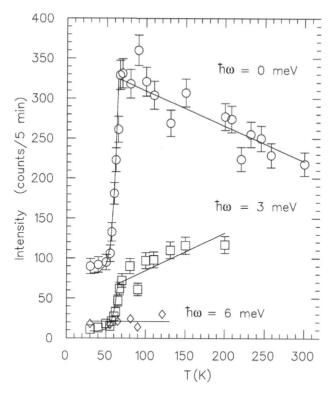

Fig. 6.10. Temperature dependence of air scattering at several energy transfers. Measurements were performed at $Q = 1.5\,\text{Å}^{-1}$ with $E_f = 14.7\,\text{meV}$.

reduced by $(T/300\,\text{K})$ following the ideal gas law. Combining the ideal gas law with vapor pressure data (and assuming a simple functional form for the temperature dependence of the elastic and inelastic cross sections) leads to the solid lines in the figure. It is important to note that the scattering is not purely elastic; there is a significant contribution at 3 meV, but it is gone by 6 meV. The inelastic N_2 scattering also has a non-monotonic Q dependence as indicated in Fig. 6.11. The non-trivial temperature and Q dependences misled us for some time when we unintentionally observed them while searching for spin fluctuations in copper oxide superconductors.

6.5.3 He

We saw above that a small amount of air in the sample can produce a surprising temperature-dependent jump in the background scattering. The attentive reader will already have realized that the same condensation phenomenon can occur with the He exchange gas. It is difficult to observe

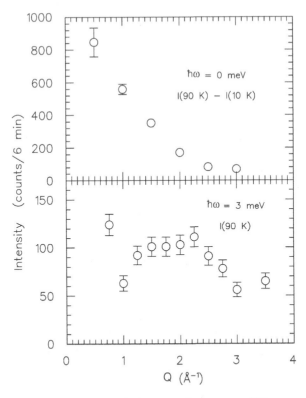

Fig. 6.11. Momentum dependence of air scattering at two different energy transfers.

because of the weak scattering cross section for ^{4}He; nevertheless, with sufficient sensitivity, it is detectable. Indeed, the He condensation effect was inadvertently observed to occur near 1.5 K by Buyers, Kjems, and Garrett (1985) in a study of the spin-fluctuation spectrum of UPt$_3$; the data are shown in Fig. 6.12. Further investigation revealed the strength of the signal to be proportional to the pressure of the He exchange gas [Buyers, Kjems, and Garrett (1986)].

A related but somewhat different manifestation of the same phenomenon was observed in a low-temperature small-angle-scattering experiment by Lynn (1986). The condensation of He onto the sample effectively smoothed the surfaces of the sample particles, thus reducing the scattering within the monitored Q range, and shifting it to much smaller Q. The transition was found to be continuous and reversible.

6.6 Techniques for spotting "spurions"

In order to chase down the cause of a spurion, one must first recognize or suspect that a particular scattering feature is not intrinsic to the sample.

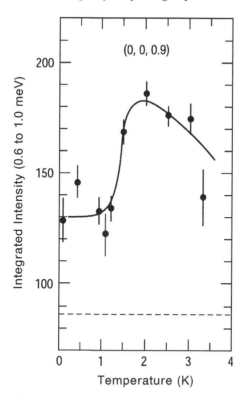

Fig. 6.12. Temperature dependence of inelastic background due to He condensation observed in a study of UPt$_3$ by Buyers, Kjems, and Garrett (1985).

One clear sign of a spurious peak is a feature that looks too beautiful! For example, a peak may occur that is much sharper than the known resolution, or much more intense than expected. In any case, once a feature is suspect, the following steps are recommended to identify the source of the signal.

(i) Begin with a quick check of the spurion flight path. By blocking the neutron beam before the sample, and then after the sample (see Fig. 6.13), one can determine whether the suspect neutrons are following the intended scattering path through the collimators, or whether they reach the detector because of shielding leakage. Leakage through gaps in the monochromator shielding is one possible source of spurious signal. Another is neutrons in the direct beam that are not intercepted by the sample; when there is insufficient shielding after the sample, the direct-beam neutrons can create a significant signal (relative to the background level) if they reach the detector shielding, when the detector happens to move into the beam path. Various

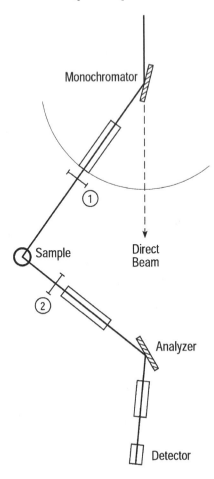

Fig. 6.13. Schematic of spectrometer, indicating positions 1 and 2 at which to block the beam to test the path of suspect neutrons.

other unexpected scattering paths are possible. Even when leakage seems unlikely, this simple test should be performed to eliminate the possibility.

(ii) Set the spectrometer to the suspect peak position, and then individually scan each of the six motors (ϕ and 2θ for monochromator, sample, and analyzer) through the nominal setting. If the peak is genuine, then moving any given motor away from its nominal position should detune the spectrometer, rapidly reducing the signal to background level. However, if, for example, the peak corresponds to an accidental Bragg scattering feature of type A (Bragg scattering at monochromator and sample, incoherent scattering at the analyzer), then the signal would be insensitive to rocking the analyzer.

(iii) Measure the suspect feature using neutrons of a different energy. This is a crucial and quite effective test. For example, if the questionable peak has been observed with a fixed final energy of 14.7 meV, then repeat the measurement with $E_f = 13.7$ meV. If, instead, one were working at $E_f = 41$ meV, try again at 39 meV. The energy change should be small enough that the resolution and scattering conditions do not change significantly, but large enough so that any dependence of the peak position on neutron energy is detectable. If the suspect feature disappears or shifts in energy, then it is likely caused by either double scattering or higher-order neutrons.

(iv) If the suspect feature passes the test above, then it must be coming from the sample or sample environment. Consider the possibility that it is coming from the sample holder (e.g., Al) or exchange gas (He or N_2). If the signal is from the sample environment, then it should not depend on the precise nature of the sample.

(v) Finally, the **Q** dependence of any intrinsic feature should be compatible with the crystal symmetry of the sample. A genuine peak should repeat in each Brillouin zone, although the intensity may depend strongly on the cross section and a possible inelastic structure factor. Also, the temperature dependence should be consistent with known properties of the sample.

References

Brockhouse, B. N., Arase, T., Caglioti, C., Rao, K. R., and Woods, A. D. B. (1962). *Phys. Rev.* **128**, 1099.

Buyers, W. J. L., Kjems, J. K., and Garrett, J. D. (1985). *Phys. Rev. Lett.* **55**, 1223.

Buyers, W. J. L., Kjems, J. K., and Garrett, J. D. (1986). *Phys. Rev. Lett.* **55**, 996E.

Currat, R. and Axe, J. D. (1978). BNL Memo G-106.

Currat, R. and Dorner, B. (1982). ILL Memo.

Ishikawa, Y., Fincher, C. R., Jr, and Shirane, G. (1980). BNL Memo G-111.

Lynn, J. W. (1986). *Physica* **136B**, 117.

Pintschovius, L., Blaschko, O., Krexner, G., de Podesta, M., and Currat, R. (1987). *Phys. Rev. B* **35**, 9330.

Shapiro, S. M., Noda, Y., Brun, T. O., Miller, J., Birnbaum, H., and Kajitani, T. (1978). *Phys. Rev. Lett.* **41**, 1051.

Shapiro, S. M., Richter, D., Noda, Y., and Birnbaum, H. (1981). *Phys. Rev. B* **23**, 1594.

Skalyo, J., Jr and Lurie, N. A. (1973). *Nucl. Instrum. Methods* **112**, 571.

Uemura, Y. J., Grier, B. H., and Shirane, G. (1982). BNL Memo G-121.

Werner, S. A. and Arif, M. (1988). *Acta Cryst. A* **44**, 383.

Werner, S. A., Giebultowicz, T. M., and Overhauser, A. W. (1987). *Physica Scripta* **T19**, 266.

7

Bragg diffraction

Although Brockhouse (1961) originally invented the triple-axis spectrometer for the study of inelastic processes, it has also proven to be an exceptionally useful instrument for measuring elastic scattering in cases where the intensity is inherently low, such as in small single crystals and powders. The three-axis instrument has played an invaluable role in studies of magnetic and structural phase transitions where new superlattice Bragg peaks appear below some critical temperature. By using an analyzer, a better signal-to-noise ratio can be obtained than with the more traditional two-axis instrument with no analyzer. The intrinsic flexibility of a three-axis instrument allows one to vary the incoming and outgoing energy, and this is invaluable in detecting and eliminating the ever-present double scattering occurring within a crystal. The triple-axis instrument has developed into a versatile powder spectrometer because of the ability to significantly change the instrumental resolution by the choice of different analyzer reflections, neutron energies, and collimations.

7.1 Three-axis vs. two-axis instruments

Using a triple-axis instrument for measuring elastic scattering can significantly improve the signal-to-noise ratio when compared to the more conventional two-axis instrument used in single-crystal diffractometers or powder spectrometers. The analyzer selects only the scattering that is elastic (within the resolution of the instrument) and the inelastic scattering from the phonons, which occurs at finite energy transfers, is automatically excluded. In a two-axis instrument this phonon scattering always appears as background, and is commonly referred to as thermal diffuse scattering (TDS). Also, since analyzer crystals such as pyrolytic graphite have reflectivities of greater than 50%, there is little intensity lost. The analyzer can be made to focus the beam onto the detector, which gives an even higher effective

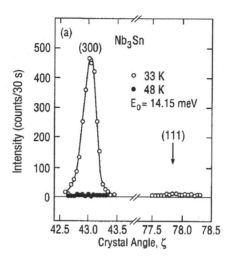

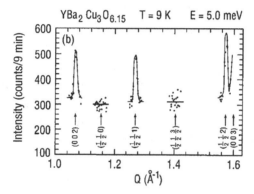

Fig. 7.1. (a) Intensities of "new" reflections in Nb_3Sn, which are forbidden in the cubic phase (after Shirane and Axe, 1971). (b) Neutron diffraction scans of several magnetic and weak nuclear peaks measured on $YBa_2Cu_3O_{6.15}$ at 9 K. The lines are guides to the eye (after Tranquada *et al.*, 1988).

reflectivity. Hence, the addition of an analyzer causes negligible intensity loss and an appreciable reduction of background. This also holds for materials that contain a significant amount of inelastic incoherent scattering which could mask weak peaks. An example is shown in Fig. 7.1(a) for the case of Nb_3Sn, which undergoes a martensitic transformation resulting in "new" peaks present below the transformation temperature. Because of the weak intensity, good signal-to-noise is necessary to observe the superlattice peaks.

The good signal-to-noise ratio achieved with a triple-axis spectrometer for elastic scattering can be effectively used in powder diffraction studies.

In these cases it is sometimes advantageous to use neutrons in the "cold" part of the reactor spectrum ($\lambda > 4.0\,\text{Å}$) instead of the thermal regime ($\lambda \sim 2.0\,\text{Å}$). With the cold neutrons, the Q-resolution is naturally better and a very high signal-to-noise ratio is obtained, even though the overall signal is significantly lower. An example is given in Fig. 7.1(b) from the study of the magnetic ordering in $YBa_2Cu_3O_{6+x}$, the prototype compound of the high-temperature "123" superconductors. The half-integer peaks are magnetic in origin and are present only at low temperatures. The increased resolution with $E_i = 5.0$ meV neutrons was necessary to separate the $(\frac{1}{2}\frac{1}{2}2)$ peak from the neighboring strong nuclear (003) Bragg peak. The measured intensity of this magnetic peak was crucial to establishing the correct magnetic structure.

The major advantage, however, of a three-axis instrument for elastic scattering is the ability and ease of changing the energy (or wavelength) of both the incoming and outgoing neutron beams. As discussed in Chap. 4, this is essential for inelastic scattering, but this capability has proven invaluable for the detection and elimination of the ever-troublesome double-scattering processes. The versatility of the three-axis instrument allows one to monitor the intensity of a Bragg peak while varying the incident and final energies with the constraint $E_i = E_f$ appropriate for elastic scattering. On the one hand, if the intensity is constant with changing energy, then the scattering process can be considered real, first-order scattering. On the other hand, if it is rapidly changing with incident energy, it is most likely double scattering. As will be discussed in §7.4, the double-scattering process is very frequently a source of erroneous results in neutron scattering studies. This is more endemic to neutron scattering when compared to X-ray scattering because larger crystals are needed for neutron studies and the relative intensity of double scattering increases with the volume. Also, the collimation in a neutron study is generally more relaxed than that used with X-rays, and this, too, increases the probability of double scattering occurring.

While descriptions and examples of double scattering make up the bulk of this chapter, there are two other topics to be covered that are relevant to elastic scattering studies. The next section discusses the proper Lorentz factor for extracting structure factors from integrated intensities of Bragg reflections, and §7.3 covers the occurrence of non-gaussian lineshapes.

7.2 Lorentz factor for a triple-axis spectrometer

In a neutron diffraction experiment involving a single-crystal sample, the measured angle-integrated intensity of a Bragg reflection is related to the structure factor of the reflection by the Lorentz factor:

$$\mathscr{I} \sim \frac{|F|^2}{\sin 2\theta_S},\tag{7.1}$$

where F is the static structure factor and the geometric prefactor, $1/\sin 2\theta_S$, is called the Lorentz factor. Eq. (7.1) is valid for the conventional situation of a double-axis instrument where a θ-2θ scan (sample and scattering angles coupled in a 1 : 2 ratio) is performed and there is no collimation in front of the counter. This formula also is valid for the case where the counter is kept fixed and the intensity is measured by rotating the sample (ϕ scan).

In using a triple-axis instrument for elastic scattering, an analyzer is added to diffract the elastically scattering neutrons, and collimation is usually added after the sample (see Fig. 4.1). In this case, the Lorentz factor for scans performed with a triple-axis instrument must be modified. Pynn (1975) analytically considered the appropriate modification of Eq. (7.1). He calculated the ratio of the integrated intensity measured in a two-axis instrument ($\mathscr{I}_2$) to that measured in a three-axis instrument ($\mathscr{I}_3$) for different types of scans and different sample mosaic widths, η_S. For a θ-2θ scan he found that, in the limit of vanishing sample mosaic width, the integrated intensity depends only upon the scattering angle through the usual $1/\sin 2\theta_S$ factor applicable for a two-axis instrument. This is shown as the horizontal line for $\eta_S = 0$ in Fig. 7.2(a), where the ratio $\mathscr{I}_2/\mathscr{I}_3$ is plotted as a function of $\sin\theta_S/\lambda$ ($= Q/4\pi$). As η_S increases there is considerable departure from the normal Lorentz factor. Figure 7.2(b) shows that for a ϕ scan there is no such simple relation between the double-axis and triple-axis intensities for any value of η_S. Fortunately, in most cases crystals with $\eta_S < 60'$ are available, so that the simple Lorentz factor of Eq. (7.1) is applicable for triple-axis measurements performed in θ-2θ mode.

Before Pynn's calculations were published, this question of the Lorentz factor was tested experimentally by Iizumi (1973) and Iizumi and Shirane (1975). They measured intensity ratios for a number of different three-axis spectrometer configurations with different collimations after the sample, as well as comparing intensities for different spectrometer configurations in a two-axis mode. They used a nearly perfect crystal of magnetite (Fe_3O_4) as the sample. Beginning with the well-established situation of a two-axis spectrometer with collimators ($20'$ in their case) before and after the monochromator and an open counter, they first tested the effect of placing collimation between the sample and counter. Figure 7.3, curve (a), shows that in θ-2θ mode adding a $20'$ collimator does not change the Lorentz factor. Measurements with a $10'$ collimator gave a similar result, albeit with a reduced intensity.

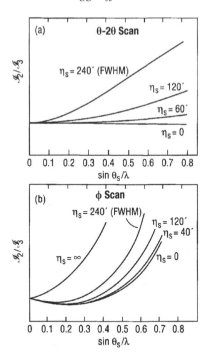

Fig. 7.2. (a) The integrated intensity of an open-detector double-axis scan ($\mathscr{I}_2$) divided by the integrated intensity of a triple-axis scan ($\mathscr{I}_3$) plotted against $\sin\theta_S/\lambda$ for a θ-2θ scan. Curves are drawn for various values of sample mosaic width η_S (with identical widths in the horizontal and vertical directions). (b) The ratios $\mathscr{I}_2/\mathscr{I}_3$ plotted against $\sin\theta_S/\lambda$ for a ϕ scan (sample rocking curve). Curves are drawn for various values of sample mosaic width η_S (from Pynn, 1975).

Conversely for ϕ scans a $10'$ collimator before the detector makes a drastic change in the Lorentz factor, as indicated by curve (b). A second set of tests compared θ-2θ scans in both three-axis and two-axis (with collimated detector) modes. The intensity ratios plotted as curves (c) and (d) in Fig. 7.3 demonstrate that, except for some reduction in intensity, the addition of an analyzer and extra collimation does not change the θ_S dependence of the measured Bragg-peak intensities.

7.3 Non-uniform lineshapes

It was shown above that one can obtain reliable integrated-intensity measurements for crystallographic studies using a triple-axis spectrometer with Soller slits and performing θ-2θ scans. It is important at this stage to point out an obvious, though frequently unappreciated fact: *one should never put two Soller slits in series without a scattering object in between.* Such a config-

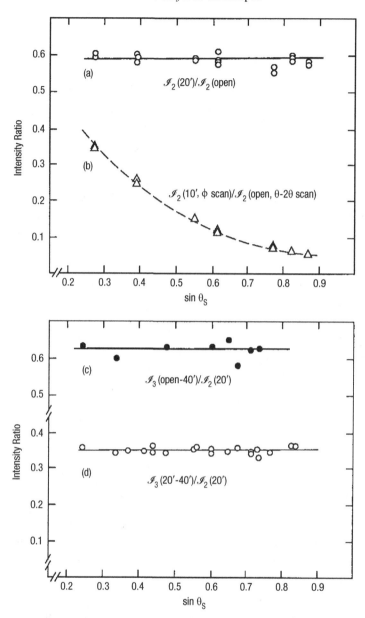

Fig. 7.3. Experimental test of the simple Lorentz factor $1/\sin 2\theta$ for integrated intensities measured in two-axis, $\mathscr{I}_2$, and three-axis, $\mathscr{I}_3$, modes. In all cases, the horizontal collimations α_0 and α_1 are equal to $20'$, with scans performed in θ-2θ mode [except as noted for curve (b)]. *Upper panel:* comparison of two-axis scans with different choices of collimation α_2 after the sample. Curve (b) compares a ϕ scan with a tight collimation, $\alpha_2 = 10'$, to a θ-2θ scan with no collimator after the sample. *Lower panel:* comparison of integrated intensities measured in three-axis mode to two-axis measurements. For each three-axis measurement, the collimations α_2-α_3 are noted in parentheses (from Iizumi and Shirane, 1975).

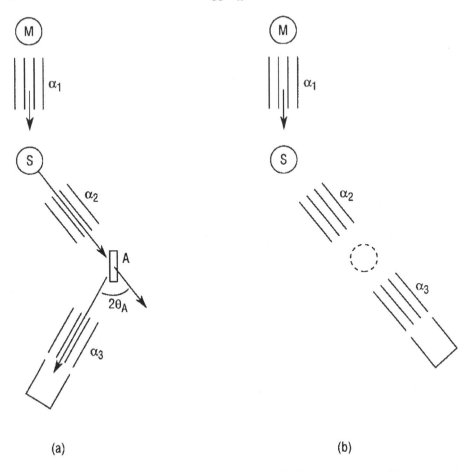

(a)   (b)

Fig. 7.4. (a) Standard three-axis arrangement. (b) Two-axis setup after conversion from three-axis.

uration can easily occur when one is converting from three-axis to two-axis mode by removing the analyzer and setting the angle of the detector arm, $2\theta_A$, to zero, as happens when aligning a sample. This situation is depicted in Fig. 7.4. Figure 7.4(a) shows a typical three-axis instrument with collimators α_2 and α_3 before and after the analyzer, respectively. In converting to the double-axis mode, $2\theta_A$ is set to zero as shown in Fig. 7.4(b) so that α_2 and α_3 become parallel, giving an effective collimation which is narrower than either collimator. An example is the case where $\alpha_2 = \alpha_3 = 20'$, resulting in a net collimation of $10'$. If there is a slight misalignment, which could easily occur due to small angular offsets, a large uncertainty in intensity could result. If the configuration shown in Fig. 7.4(b) must be used, one should assure that α_3 is considerably larger than α_2 (or vice versa).

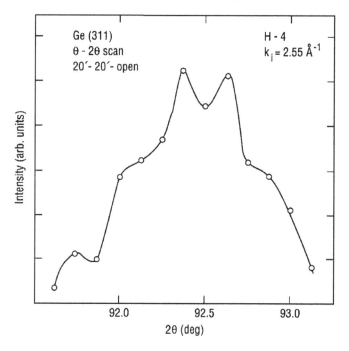

Fig. 7.5. Example of non-gaussian lineshape observed by Moncton, Lynn, and Shirane (1974) at high scattering angles.

Another type of problem is sometimes encountered in normal three-axis operation. The approximations used in the calculation of the resolution function (Chap. 4) yield gaussian lineshapes for all scan types. Thus, the profile of a θ-2θ scan of a single crystal should yield a gaussian lineshape. However, it was noted some years ago [Moncton, Lynn, and Shirane (1974)] that in certain cases a multi-peaked, non-gaussian lineshape, like the one shown in Fig. 7.5, can result if the sample size (or beam size) is comparable to or smaller than the distance between the blades of the collimator α_0 between reactor and monochromator. This size is typically 3 mm, and the effect is most pronounced at large scattering angles $2\theta_S > 90°$.

The origin of this effect was explained by Kjems and Satija (1981) as a shadowing of the crystal by the collimator blades of the collimator α_0 as shown in Fig. 7.6. Figure 7.6(b) shows an expanded view of the collimator. The intensity distribution $I(x, y)$, where x is measured perpendicular to the axis of the collimator and y is the angle of observation relative to the axis of the collimator, is shown in Fig. 7.6(c). The intensity is constant within the shaded areas of the curve, so that at a fixed x the distribution as a function of y has a pronounced peak structure, with an angular splitting between the

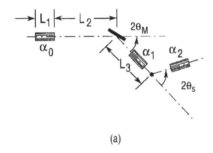

(a)

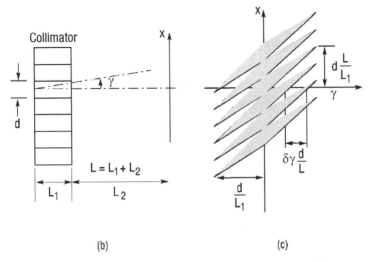

(b)   (c)

Fig. 7.6. Analysis of non-gaussian lineshape problem. (a) Monochromator and sample arms of a three-axis spectrometer, with the dimensions L_1, L_2, and L_3 indicated. (b) Distorted view of the α_0 Soller collimator, defining the blade spacing d, the beam divergence angle γ, and the transverse position coordinate x. (c) Schematic diagram of the intensity distribution at the sample position in the γ–x phase space (from Kjems and Satija, 1981).

the peaks of

$$\delta\gamma = d/L, \tag{7.2}$$

where d is the spacing between the blades and L is the distance from the collimator entrance to the monochromator position. If a monochromator is placed at a distance L_2 from the Soller slit collimator, the angular variation of intensity will be transformed into a modulation of the wave vector, Δk, of the diffracted neutrons by Bragg's Law:

$$\Delta k = \frac{\delta\gamma}{2} k \cot \theta_M, \tag{7.3}$$

where θ_M is the Bragg angle for the monochromator reflection. If a sample is now placed after the monochromator, structure will show up in a θ-2θ scan with an angular separation given by

$$\Delta\theta_S = (\Delta k/k)\tan\theta_S. \tag{7.4}$$

Combining Eqs. (7.2)–(7.4), one finds that the observed peak-splitting should be given by

$$\begin{aligned}\Delta(2\theta_S) &= (d/L)\tan\theta_S\cot\theta_M, \\ &= \alpha_0(L_1/L)\tan\theta_S\cot\theta_M, \tag{7.5}\end{aligned}$$

where, in the second line, we have made use of the fact that $\alpha_0 = d/L_1$. Because of the factor $\tan\theta_S$ in Eq. (7.5), the splitting will increase rapidly for $2\theta_S > 90°$. For a point sample placed at a distance $L_3 \sim L$ from the monochromator, the structure will appear independent of the collimation between the monochromator and sample due to focusing. The splitting is only dependent upon the collimator α_0.

This effect was tested at the H8 instrument at the HFBR using a nearly perfect crystal of germanium with a known gaussian mosaic distribution. The dimensions of the instrument give $L_1 = 90\,\text{cm}$, $L_2 = 180\,\text{cm}$, $2\theta_M = 41°$ and, for the Ge (331) reflection, $2\theta_S = 130°$. Putting these into Eq. (7.5) yields

$$\delta(2\theta_S)_{\text{calc}} = 1.9 \times \alpha_0. \tag{7.6}$$

The experimental results are shown in Fig. 7.7 where different beam widths are used to approximate different sample sizes. The observed splitting, when converted to angle, corresponds to

$$\Delta(2\theta_S)_{\text{obs}} = 2.1 \times \alpha_0, \tag{7.7}$$

which is in reasonable agreement with Eq. (7.6). Figure 7.7 demonstrates that the effect becomes more pronounced with decreasing beam width. For the larger beam width, or larger sample size, the effect of the modulation is washed out and is not as apparent.

7.4 Double scattering

We turn now to the problem of double scattering. More precisely, the phenomenon with which we are concerned should be called multiple scattering. A detailed analysis of multiple-scattering effects when two reflections occur simultaneously has been given by Moon and Shull (1964). We will be particularly interested in the case in which a Bragg peak can appear at a forbidden

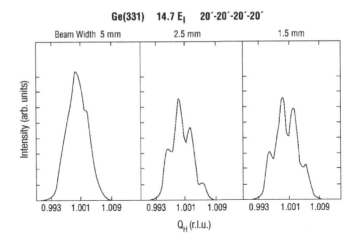

Fig. 7.7. Examples of diffraction line profiles, measured at a large scattering angle, as a function of beam width (effective sample size) as discussed in the text (after Kjems and Satija, 1981). [r.l.u. = reciprocal-lattice units.]

position. Such a peak is sometimes called a "Renninger reflection" after the scientist who first observed this effect with X-rays (Renninger, 1937); nevertheless, we shall follow convention and use the term double scattering in the remainder of this section.

Double scattering has been, and continues to be, the cause of numerous misinterpretations published in the scientific literature. Since double-scattering peaks frequently appear at positions where a proper Bragg reflection is forbidden, they can easily be mistaken for a superlattice peak associated with a structural phase transition or antiferromagnetic order. Fortunately, using a triple-axis instrument it is relatively straightforward to test a suspect peak for double scattering.

To describe the double-scattering condition, we consider the Ewald sphere (see Fig. 7.8) of radius $k = k_i = k_f$, with the tip of the incident wave vector $\mathbf{k}_i$ coinciding with the origin of reciprocal space (point 0). The Laue condition will be satisfied when a reciprocal-lattice point $\mathbf{G}$ lies on the surface of the sphere. A triple-axis spectrometer detects the scattered intensity at momentum transfer $\mathbf{Q}_0$ in the scattering plane defined by the two wave vectors $\mathbf{k}_i$ and $\mathbf{k}_f$. If $\mathbf{Q}_0$ coincides with a reciprocal-lattice vector $\mathbf{G}_0$, a Bragg peak is detected. Figure 7.8 shows a situation in which a second reciprocal-lattice point, $\mathbf{G}_1$, also lies on the Ewald sphere. (Note that the second reflection need not lie in the scattering plane.) The simultaneous reflection can affect the intensity reaching the detector in two ways: (a) the measured intensity can be lowered due to depletion of the incident beam by

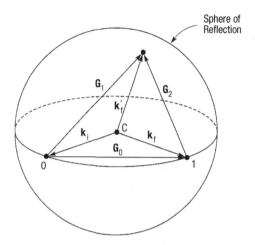

Fig. 7.8. Geometric representation of simultaneous reflections in the reciprocal lattice. The origin (0) and two reciprocal-lattice points defined by the vectors G_0 and G_1 lie on the Ewald sphere. The wave vectors k_i and k_f denote the initial and final beams defined by the spectromer; k'_f is a secondary beam (from Moon and Shull, 1964).

the second reflection, and (b) it can be increased if neutrons diffracted by G_1 undergo a second scattering by a reciprocal-lattice vector $G_2 = G_0 - G_1$. Conversely, it is possible for neutrons in the beam scattered along k_f to be lost to the k'_f beam by a second scattering through $-G_2$. In general, the net gain or loss of intensity to the detected beam will depend in a complicated way on the sample geometry (Moon and Shull, 1964). The situation simplifies when the intensity of the reflection G_0 is very weak or zero; in that case, the dominant effect is the gain in intensity due to the double scattering through $G_1 + G_2$.

The stringent condition for double scattering is that more than one reciprocal-lattice point must lie on the surface of the Ewald sphere. If the Ewald sphere were truly an infinitesimally thin shell, then one might expect the probability of having simultaneous reflections to be rather low. However, there are several factors that tend to make double scattering particularly problematic in neutron scattering experiments:

(i) The experimentally-determined Ewald sphere should be thought of as a shell with a significant thickness Δk resulting from the spreads Δk_i and Δk_f in the incident and final wave vectors allowed by the finite divergence of the neutron beams and the mosaic natures of the monochromator and analyzer. A coarse $\Delta k/k$ substantially enhances the probability of having two reciprocal-lattice vectors "touching" the Ewald sphere. Of course, collecting a large solid angle of the neutron beam and focusing it onto a small sample area will further exacerbate this problem. In contrast, double scattering tends to be much less of a problem in X-ray diffraction, where the greater source intensity allows one to work with a much smaller $\Delta k/k$.

(ii) The probability of double scattering depends on the area of the Ewald sphere, which varies as $\sim k^2$ (λ^{-2}). Thus, one can lower the chances of double-scattering contamination by utilizing the smallest possible wave vector. Cold neutrons, with $k_i < 1.55\,\text{Å}^{-1}$ ($E_i < 5.2\,\text{meV}$ or $\lambda > 4\,\text{Å}$), are often quite suitable for this purpose. This experimental window can be particularly clean, since the use of a cold Be filter (with a 5.2-meV cutoff) effectively eliminates any higher-order harmonic neutrons from the incident beam.

(iii) Given that a double-scattering condition exists, the intensity of the contaminating signal is proportional to the square of the product of the elastic structure factors, $F(\mathbf{G}_1)$ and $F(\mathbf{G}_2)$, and the volume of the sample. In an X-ray experiment, the relatively large absorption coefficient tends to limit the effective sample volume, thus minimizing the double-scattering problem. In a neutron-scattering experiment, on the other hand, the typically much larger sample volume makes double scattering a much more likely occurrence. This is one of the reasons (extinction being the other) that crystallographers work with quite small samples.

Contamination of weak reflections by double scattering is so pervasive that it is good practice to assume that it is present until tests prove otherwise. There are two methods to test for the presence of double scattering. The first method is to rotate the crystal about the scattering vector, $\mathbf{Q}$, thought to be contaminated. This rotates the crystal relative to the Ewald sphere, causing possible secondary reflections to sweep through the sphere while maintaining the primary Bragg condition. Moon and Shull (1964) demonstrated this method using a single crystal of iron. Figure 7.9 shows the intensity of the allowed $\mathbf{G}_0 = (200)$ reflection of iron as a function of the rotation angle of the sample about $\mathbf{G}$, a so-called azimuthal scan, measured both in reflection (upper curve) and transmission (lower curve) geometries. The dotted line represents the intensity which is free of double scattering. The peaks and valleys correspond to double-scattering conditions, and the $\mathbf{G}_1$ and $\mathbf{G}_2$ which contribute to each process are indicated in the figure. These measurements demonstrate that both increases and decreases of the observed intensity relative to the uncontaminated level may occur, depending upon the relative importance of beam depletion versus the additional scattering. As shown here, the effect can easily reach 50% of the "true" intensity.

An azimuthal scan is generally impractical for a crystal mounted in a cryostat, furnace, or magnet. For such cases, there is a second method which is particularly convenient to apply on a triple-axis instrument. The idea

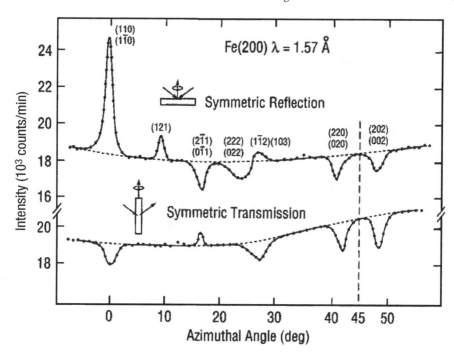

Fig. 7.9. Simultaneous reflection effects observed in the (200) reflection from iron as the crystal is rotated around the scattering vector. Pronounced differences are to be noted for the reflecting and transmitting configurations (from Moon and Shull, 1964).

is to change the radius of the Ewald sphere by simultaneously scanning k_i and k_f while maintaining the Bragg condition for $\mathbf{Q}_0$. As the radius of the Ewald sphere is varied, its surface sweeps through reciprocal space and through possible secondary reflections. The procedure is to monitor the elastic intensity measured at $\mathbf{Q}_0$ as a function of the incident energy, E_i. Double scattering reveals itself by a rapid variation of intensity with E_i, while for a "true" signal the intensity is nearly constant as a function of E_i. Note that this test is impossible to perform with conventional X-ray sources since E_i is fixed, although it is feasible with synchrotron-produced X-rays.

In the next three sections we will present a number of examples in which the study of structural or magnetic phase transitions is complicated by double scattering; however, before ending this section we consider another case in which weak reflections are of interest. In certain crystal stuctures, scattered intensity can appear at otherwise forbidden reflections due to anharmonic lattice vibrations or non-spherical electronic charge distributions (the latter contribution, of course, being detectable only with X-rays). Such is the case for the strongly-covalent semiconductors Ge and Si. Both of these materials crystallize in the diamond structure, space group $Fd3m$, which consists of two interpenetrating face-centered-cubic lattices. The conditions for Bragg scattering are:

$$h + k + l = 4n \quad \text{or} \quad 2n + 1.$$

Thus, no intensity should be present at the $\mathbf{G}_0 = (2, 2, 2)$ position. Double scattering from two allowed reflections such as $\mathbf{G}_1 = (3, 3, 3)$ and $\mathbf{G}_2 = (1, 1, 1)$, which each have a large structure factor, can add to give $\mathbf{G}_0 = (2, 2, 2)$. In a neutron diffraction study on Si by Keating *et al.* (1971), careful analysis of the double-scattering conditions led to proper determinations of the temperature-dependent intensity at (222) due to cubic anharmonicity. Hastings and Batterman (1975) later succeeded in detecting the (442) and (622) reflections. The temperature dependence of the forbidden (442) reflection was also studied by Trucano and Batterman (1972) using X-ray diffraction. By making use of the neutron result for the (222) reflection, they were able to determine the contribution due to the antisymmetric part of the electronic distribution associated with the covalent bonds.

7.5 Double scattering in structural phase transitions

The study of structural phase transitions continues to be one of the major topics in solid-state physics. The soft-mode theory was developed simultaneously by Cochran (1960) and Anderson (1960) to describe the origin of ferroelectricity, and it has since been successfully applied to explain a wide variety of structural transformations. In essence, this theory states that the eigenvector of a certain lattice vibrational mode describes the pattern of atomic displacements necessary to transform the solid from its high-temperature phase into the low-temperature phase. As temperature is reduced towards the transition temperature, T_c, the frequency of this mode decreases, eventually vanishing at T_c and resulting in the structural transformation. The transition can be viewed as a "freezing" of the relevant vibrational mode. Both inelastic neutron scattering and light scattering have been successfully employed to characterize the soft-mode dynamics in a number of ferroelectrics such as $BaTiO_3$, $KTaO_3$, $PbTiO_3$, $KNbO_3$, where the soft mode occurs at $q \sim 0$. In the case of antiferroelectrics, the soft mode is located at the zone boundary, a region of reciprocal space that can only be probed directly by neutron scattering. Detailed studies have been performed on many of the perovskites, such as $SrTiO_3$, $KMnF_3$, $CsPbCl_3$, and $LaAlO_3$. When a mode freezes out at a zone-boundary point, a new Bragg peak generally appears at that position in reciprocal space. The intensity of this superlattice peak is proportional to the square of the order parameter and, consequently, it increases as the temperature is reduced. A study of the intensity of these new peaks is important in establishing the low-temperature structure, so care must be taken to eliminate contamination due to multiple scattering. Below we discuss the cases of the antiferroelectric transition in $SrTiO_3$, and the martensitic transition in the superconductor Nb_3Sn.

7.5.1 SrTiO₃

The structural phase transition occurring at $T_c = 110\,K$ in $SrTiO_3$ is a typical example of a soft-mode phase transition where the soft-mode wave vector occurs at a zone-boundary point. The high-temperature phase of $SrTiO_3$ has the cubic perovskite structure (space group $Pm3m$). Below $110\,K$, $SrTiO_3$ transforms to a tetragonal structure, $I4/mcm$, with an extremely small c/a ratio of 1.0005. The transition is most easily detected by the appearance below T_c of new superlattice peaks at the zone-boundary positions $(\frac{h}{2}\frac{k}{2}\frac{l}{2})$ of the cubic phase, where h, k, and l are all odd. The transition has been extensively studied by electron spin resonance, Raman scattering, and neutron scattering (Shirane and Yamada, 1969; Cowley, Buyers, and Dolling, 1969). It was established that the transition is the result of a soft mode with symmetry Γ_{25} at the R-point, $(\frac{1}{2}\frac{1}{2}\frac{1}{2})$, of the Brillouin zone. This mode corresponds to a rotation of the oxygen octahedra about one of the three cubic axes. If this is the correct description of the low-temperature structure, then certain $\{\frac{h}{2}\frac{k}{2}\frac{l}{2}\}$ reflections are not allowed by symmetry, namely $\{\frac{h}{2}\frac{h}{2}\frac{h}{2}\}$. The original neutron study of Shirane and Yamada (1969) observed a weak intensity at $(\frac{1}{2}\frac{1}{2}\frac{1}{2})$ but suggested that it could be due to double scattering. In a later study, Okazaki and Willis (1980) claimed that the intensity at $(\frac{1}{2}\frac{1}{2}\frac{1}{2})$ was real because it had a strong temperature dependence and followed that of the order parameter. If the latter claim were true, then the generally accepted soft-mode model for $SrTiO_3$ would have to be modified.

A subsequent experiment was performed in order to prove that the observed intensity is indeed due to double scattering (Shirane *et al.*, 1981). Figure 7.10 shows the intensity of several superlattice peaks measured as a function of neutron energy. The $(\frac{3}{2}\frac{1}{2}\frac{1}{2})$ peak whose intensity is plotted in Fig. 7.10(a) is an allowed superlattice reflection. Although its intensity is only 3% of the fundamental (200) Bragg peak, it remains essentially constant over the entire energy range. Thus, any double scattering is only a small fraction of the real signal. In Fig. 7.10(b) two forbidden reflections are compared in the same manner. For the $(\frac{3}{2}\frac{3}{2}\frac{3}{2})$ reflection there is a large relative variation

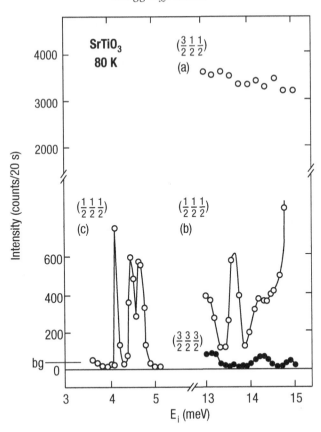

Fig. 7.10. Double scattering at the forbidden $(\frac{1}{2}\frac{1}{2}\frac{1}{2})$ point of SrTiO$_3$. The strong allowed superlattice reflection $(\frac{3}{2}\frac{1}{2}\frac{1}{2})$ is almost free from double scattering (a), while (b) and (c) show the unusually strong double scattering for $(\frac{1}{2}\frac{1}{2}\frac{1}{2})$. Note that the observed intensity reaches the background level (bg) only near 4 meV. The filled symbols in (b) show the intensity of another forbidden reflection, $(\frac{3}{2}\frac{3}{2}\frac{3}{2})$ (from Shirane *et al.*, 1981).

of intensity with energy over the range of 13–15 meV; however, the minimum intensity observed near $E_i = 13.5$ meV coincides with the background level, thus indicating the lack of any intrinsic signal at this wave vector. The data for the other forbidden peak, $(\frac{1}{2}\frac{1}{2}\frac{1}{2})$, show a much more dramatic variation of intensity with energy, with the signal always staying above the background level. One might conclude from this that there was some true signal present. However, in another experiment [Fig. 7.10(c)] the incident energy was lowered to 5 meV, thereby reducing the size of the Ewald sphere and with it the probability of double scattering. Figure 7.10(c) shows that while there is still some contamination at $E_i = 4.1$, 4.5 and 4.7 meV, the intensity level for $E_i < 4.0\,\text{meV}$ corresponds to the background. Using the "minimum principle", one concludes that the peaks such as $\{\frac{h}{2}\frac{h}{2}\frac{h}{2}\}$ are forbidden, and that the structure determined from the soft-mode symmetry is valid.

In the experiment of Okazaki and Willis (1980), neutrons with $\lambda = 1.0\,\text{Å}$ ($E = 80\,\text{meV}$) were used, giving an Ewald sphere with a very large radius. The temperature dependence they observed is the result of a double scattering process involving a fundamental reflection, $\mathbf{G}_1 = (h, k, l)$, and a half-integer superlattice reflection, $\mathbf{G}_2 = (\frac{h'}{2}, \frac{k'}{2}, \frac{l'}{2})$. The intensity variation at $\mathbf{Q} = \mathbf{G}_1 + \mathbf{G}_2$ mirrored the temperature-dependent behavior of the half-integer peak, $\mathbf{G}_2$. The double scattering in $SrTiO_3$ is one of the most persistent cases of this very common phenomenon in all of neutron scattering!

7.5.2 Nb_3Sn

Nb_3Sn is another interesting example of a structural phase transition in which the elimination of double scattering was crucial to an understanding of the transition and the low-temperature structure. Nb_3Sn is a superconductor which, when first discovered, had one of the highest known transition temperatures, $T_c = 18\,\text{K}$. At a somewhat higher temperature ($T_m = 45\,\text{K}$) it undergoes a martensitic transformation, and it was suspected that the structural and superconducting transitions might be related. For $T > T_m$, the symmetry is cubic with space group $Pm3m$ (O_h^3) with certain reflections, namely (hhl) with $l = 2n+1$, not allowed. As T approaches T_m, a remarkable softening of the elastic constant $c' = \frac{1}{2}(c_{11} - c_{12})$ occurs, which is associated with a particular shear mode propagating along [110] with polarization along [1$\bar{1}$0]. Below T_m, the system becomes tetragonal with $c/a = 1.006$. Figure 7.11 shows four different proposed models of atomic displacements in the martensitic phase. The correct model can be decided by establishing which of several possible new Bragg peaks, in particular the (111), (300), or (003), is present in the low-temperature tetragonal phase. These reciprocal-lattice points can be heavily contaminated by double scattering because two allowed Bragg peaks of the cubic phase can be combined to give the Miller indices of these reflections. The approach to minimizing the double scattering is to find the contamination-free energy window in the high-temperature phase, where the Bragg peaks of interest are forbidden. Figure 7.12 shows the intensity at the (111) and the (300) positions as a function of energy in the range 13.5–15 meV measured at $T > T_m$ (i.e., in the cubic phase). Since these Bragg peaks are known to be absent, the intensity should be zero. It is clear in Fig. 7.12 that (111) is contaminated at both ends of the energy range scanned while the (300) position is free of contamination. The appro-

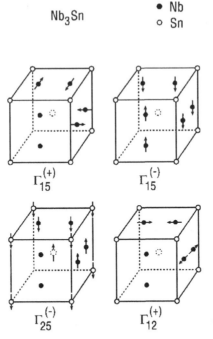

Fig. 7.11. Four models for the atomic displacements in the low-temperature tetragonal martensitic phase of Nb_3Sn (from Shirane and Axe, 1971).

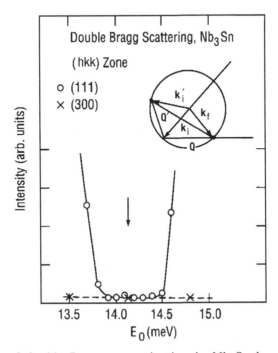

Fig. 7.12. Study of double Bragg contamination in Nb_3Sn by changing neutron energy E_0. The (300) is clean, while the (111) is heavily contaminated at both ends (from Shirane and Axe, 1971).

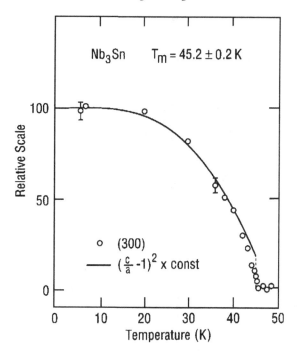

Fig. 7.13. Temperature dependence of the (300) reflection in Nb_3Sn. Solid line corresponds to the spontaneous tetragonal strain. Strictly speaking, the abscissa is $\alpha(T/T_m)$ with $\alpha = 45.2$. This was done to compensate for the variation in T_m between the samples compared (from Shirane and Axe, 1971).

priate contamination-free energy chosen was $E = 14.2$ meV. On cooling, an increase in intensity was observed at (300), with no intensity at (003) and negligible signal at (111). This observation was sufficient to decide that the $\Gamma_{12}^{(+)}$ mode (Fig. 7.11) is correct for the atomic displacements associated with the low-temperature phase. The correct space group is then $P4_2/mmc$ (D_{4h}^2), for which the forbidden reflections are of the type (hhl) with $l = 2n + 1$. [Note that this extinction rule appears similar to that for the cubic phase. The difference is that in tetragonal symmetry, (hhl) is no longer equivalent to (lhh) or (hlh). Thus, (300) is allowed but (003) and (111) are not.] The temperature dependence of the order parameter is given by the intensity of the new reflection at (300) as shown in Fig. 7.13.

7.6 Double scattering in magnetic transitions

Double scattering can also plague the study of magnetic transitions. In antiferromagnets there is usually a doubling in the size of the unit cells in at least one direction and consequently magnetic superlattice peaks appear

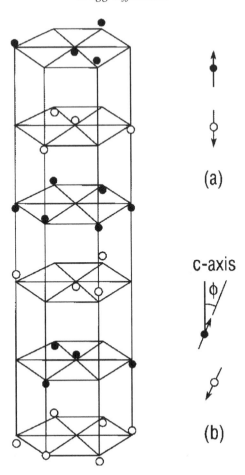

Fig. 7.14. Magnetic structure of $FeTiO_3$. Filled and open circles denote + and − spins, respectively. Spins are restricted to the c-axis in model (a), and are canted from the c-axis by angle ϕ in model (b) (from Yamaguchi *et al.*, 1986).

with half-integer values of the Miller indices. Thus, a pure nuclear peak and a magnetic peak can combine by double scattering to yield a half-integer peak at some forbidden position.

7.6.1 Fe TiO₃

Ilmenite, $FeTiO_3$, is an antiferromagnet with the magnetic moments of adjacent hexagonal planes coupled antiparallel to each other as shown in Fig. 7.14 (Yamaguchi *et al.*, 1986). The large magnetic anisotropy restricts the moment to being directed along the c-axis. There has been some controversy as to whether the moment lies exactly parallel to the c-axis ($\phi = 0$), or

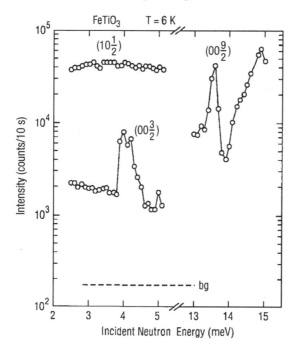

Fig. 7.15. Bragg-peak intensities measured from $FeTiO_3$ as a function of the incident neutron energy. The intensity of the $(00\frac{3}{2})$ and $(00\frac{9}{2})$ reflections is shown for $2.5 < E_i < 5.5\,meV$, and the crystal size effect is shown for $13 < E_i < 15\,meV$. The flat regions between sharp peaks give the true intensity (from Yamaguchi *et al.*, 1986).

whether there is a small amount of canting ($\phi \neq 0$) as depicted in Fig. 7.14. If $\phi = 0$, then all $(00\frac{l}{2})$ reflections should vanish. In a neutron diffraction study, the observed intensity at $(00\frac{3}{2})$ led to the conclusion that the canting angle was $\phi = 2.2°$. Since the intensity of this peak was weak, there was a possibility that it arose from the combination of a (hkl) nuclear and a $(\frac{h'}{2}\frac{k'}{2}\frac{l'}{2})$ magnetic peak. A test was performed by monitoring several of the $(00\frac{l}{2})$ magnetic reflections while varying the incident and final energy; the results are shown in Fig. 7.15. In the 13–15-meV range there is severe contamination as shown by the large change in signal with energy. (Note that the scale is logarithmic). The measurements in the range $E_i \leq 5$ meV also show sharp peaks, but a flat region in between the peaks establishes that there is indeed a true signal at $(00\frac{l}{2})$. The refined value for ϕ is $(1.6 \pm 0.1)°$. Figure 7.15 also shows that the intensity at $(10\frac{1}{2})$, a strong allowed magnetic peak, is uniform with energy and free from double-scattering contamination.

Another way in which to obtain superlattice intensities that is relatively free of multiple scattering is to use a polycrystalline sample for the diffraction

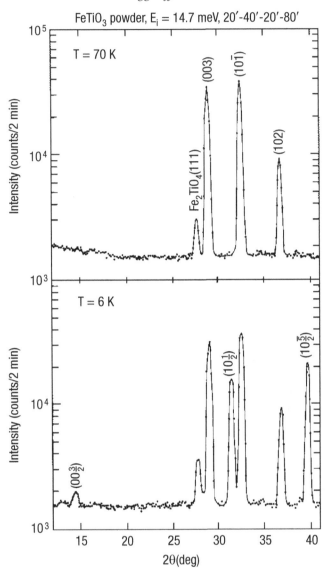

Fig. 7.16. Powder diffraction pattern of FeTiO$_3$ (from Yamaguchi *et al.*, 1986).

measurement. As long as the crystallites are small and randomly oriented, the probability of contamination by double scattering is negligible. Of course, one drawback to using powder diffraction to study very weak peaks is that the signal-to-background ratio is much lower than for a single-crystal measurement. In the case of FeTiO$_3$ this is not a problem. The powder diffraction pattern shown in Fig. 7.16 clearly reveals the presence of the $(00\frac{3}{2})$ magnetic Bragg peak below the Néel temperature, T_N, which is 68 K.

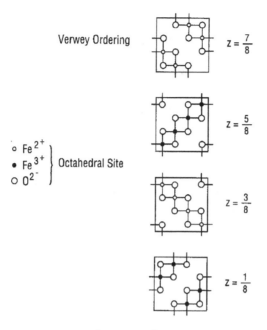

Fig. 7.17. Ionic arrangement of Fe^{2+} and Fe^{3+} on B-sites in Fe_3O_4 due to Verwey ordering (from Shirane *et al.*, 1975).

7.6.2 Fe₃O₄

Magnetite, Fe_3O_4, is an example of an "inverted" spinel structure which crystallizes in a face-centered-cubic structure with space group $Fd3m$ (O_h^7). There are 8 formula units per unit cell, with 8 tetrahedral A sites occupied solely by Fe^{3+} and 16 octahedral B sites occupied randomly by equal numbers of Fe^{2+} and Fe^{3+} ions. The conditions for allowed reflections are that h, k, and l are either all odd or all even due to the face centering, with the additional condition that for $(0kl)$, $k + l = 4n$. Below $T_N = 850\,\mathrm{K}$, the Fe moments on the A- and B-sites order antiparallel to one another. The net magnetic order is ferrimagnetic because of the unequal numbers and magnitudes of moments on the two sets of sites. At about the so-called Verwey temperature $T_V = 120\,\mathrm{K}$, magnetite undergoes another phase transition that was first identified by resistivity, specific heat, and magnetic studies. Verwey and Haayman (1941) proposed that this transition was due to the ordering of Fe^{2+} and Fe^{3+} atoms on the octahedral sites (Fig. 7.17) causing a reduction in symmetry. In a single-crystal diffraction study, Hamilton (1958) measured a finite intensity at the forbidden (002) position that apparently confirmed the model proposed by Verwey.

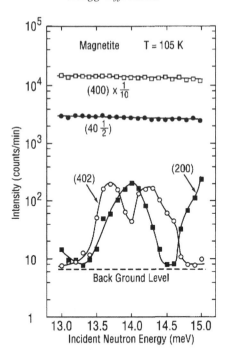

Fig. 7.18. Peak intensities measured on Fe_3O_4 as a function of the incident neutron energy. The fluctuations in the (402) and (200) intensities indicate the presence of multiple scattering (from Shirane *et al.*, 1975).

It was later shown by Shirane *et al.* (1975) that the ubiquitous double-scattering problem was responsible for the intensity that Hamilton observed. This is illustrated in Fig. 7.18, where the intensities of the forbidden (402) and (002) peaks are shown as a function of incident energy at $T = 105\,K < T_V$. The now familiar fluctuations of intensity with incident energy provide evidence of double scattering. In contrast, the intensities of two allowed Bragg peaks, the (400) and $(40\frac{1}{2})$, are nearly constant with energy. Figure 7.19 demonstrates a second test for multiple scattering, an azimuthal scan of the crystal about the scattering vector. The intensity of the (200) peak is strongly modulated, whereas the intensity of the allowed (400) peak is nearly constant with angle. The true structure of the low-temperature phase of magnetite was finally decided by the experiment of Iizumi *et al.* (1982) in which extreme care was taken to minimize the double scattering.

7.6.3 *MnF₂*

MnF_2 is a somewhat unusual example of an antiferromagnet in which the chemical and magnetic unit cells are identical. The magnetic order is iden-

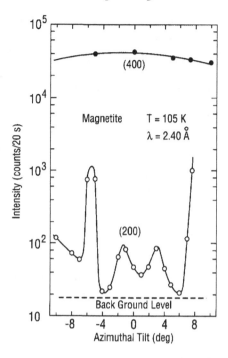

Fig. 7.19. Peak intensities measured on Fe_3O_4 as a function of the azimuthal tilt about the scattering vector. The oscillations of the (200) intensity demonstrate the presence of multiple scattering (from Shirane *et al.*, 1975).

tified by the appearance below T_N of new Bragg peaks at positions that are forbidden for nuclear scattering. To understand this, we first observe that MnF_2 has the tetragonal rutile structure, with space group $P4_2/mnm$ (D_{4h}^{14}). The magnetic Mn atoms occupy the corners and the body center of the unit cell, while the four F atoms are distributed as shown in Fig. 5.14. The only special condition for allowed reflections is that for $(h0l)$, $h + l = 2n$. The original powder diffraction measurements by Erickson (1953) clearly showed, for $T < T_N$, the presence of magnetic peaks at the nuclear-forbidden positions (100) and (201), as well as additional intensity at the allowed (111) and (210) positions. These results, when combined with the absence of any signal at (001), led uniquely to the antiferromagnetic structure shown in Fig. 5.14, with the spins pointing along the c-axis.

Because the magnetic moment of Mn^{2+} is quite large ($\sim 5\mu_B$), the magnetic scattering measured by neutron diffraction is sufficiently strong that contamination of magnetic Bragg peaks by double scattering (which can be achieved by two nuclear reflections) is not a significant problem. In contrast, the magnetic intensity obtained by X-ray diffraction is only on the order of 10^{-6} compared to the charge scattering at allowed reflections of the chemical cell. At this level, double scattering can become a significant problem even when using a high-resolution synchrotron X-ray spectrometer, as was found in a magnetic X-ray diffraction study of MnF_2 by Goldman *et al.* (1987). In this particular case, it was possible to eliminate the double-scattering condition by rotating the crystal about the scattering vector; however, experience has shown that there are some magnetic systems for which no solution to the X-ray double-scattering problem has been found. This difficulty appears to be one area in which neutrons retain a distinct advantage.

7.7 Double scattering from twins

In all of the examples discussed above, peaks due to double scattering appeared at positions **Q** commensurate with the reciprocal lattice. In twinned crystals, it is possible for double scattering to result in apparently incommensurate peaks. This variant was first noticed in KDP (potassium di-hydrogen phosphate, KH_2PO_4)-type ferroelectrics. Specifically, neutron diffraction measurements on the ferroelectric phase of CsD_2AsO_4(CDA) by Dietrich, Cowley, and Shapiro (1974) exhibited incommensurate peaks with positions corresponding to periodicities of 30, 15, 10 and 5 lattice spacings. It was subsequently shown by Meyer *et al.* (1975) that the extra peaks result from double Bragg scattering involving two different ferroelectric domains.

A recent study (Chou, Schmidt, and Shapiro) of the mixed antiferro-electric–ferroelectric system $Rb_{0.9}(ND_4)_{0.1}D_2AsO_4$ (DRADA) found similar features. This crystal undergoes a ferroelectric phase transition at 135 K accompanied by a change of crystal symmetry from tetragonal (space group $I\bar{4}2d$) to orthorhombic (space group $Fdd2$). The orthorhombic cell can be described as a distortion of the tetragonal cell by a spontaneous (100) or (010) shear. Twin domains result from four equivalent shears in the basal plane. In the (001) zone, a tetragonal Bragg peak at ($hk0$) is split into ($h \pm k\delta, k \pm h\delta, 0$) as indicated in Fig. 7.20(a) (not to scale), where $\delta = \tan\chi$, with χ being the shear angle in the ferroelectric phase. Figure 7.20(b) shows a transverse scan through (200) measured in the ferroelectric phase, which clearly shows the splitting due to domains. The peaks appear at $k = 2\delta = 0.0275$ which corresponds to a shear angle of $\pm 0.8°$. In addition to these ferroelectric Bragg peaks at $(2, \pm 2\delta, 0)$, a longitudinal scan reveals more peaks at $(2 \pm 4\delta, 0, 0)$ and $(2 \pm 8\delta, 0, 0)$, as shown in Fig. 7.20(c) and shown schematically as open circles in the reciprocal-space diagram. These longitudinal peaks could

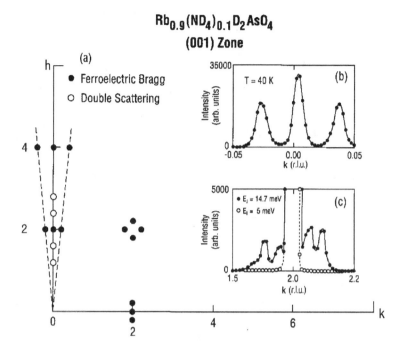

Fig. 7.20. Double scattering from a twinned crystal of $Rb_{0.9}(ND_4)_{0.1}D_2AsO_4$.

easily be interpreted as evidence of a modulated structure, as was done in the paper on CDA (Dietrich, Cowley, and Shapiro 1974); however, closer examination shows that they arise from double scattering. This is proven by repeating the scan at a different incident energy. Figure 7.20(c) shows that, using $E_i = 5.0$ meV, the peaks have completely disappeared! A specific example may clarify how they can result from double scattering. Consider the spurious $(2 \pm 4\delta, 0, 0)$ longitudinal peaks. These could arise from the ferroelectric Bragg reflections present about (220) and $(0\bar{2}0)$, that is,

$$(2 \pm 4\delta, 0, 0) = (2 \pm 2\delta, 2, 0) + (\pm 2\delta, -2, 0). \tag{7.8}$$

7.8 Varying resolution by choice of analyzer

High-resolution neutron powder diffraction studies have advanced a great deal in recent years. A major contribution to this advancement is the development of the profile-refinement technique by Rietveld (1969). In this technique, each data point is used to compare with an assumed model of the structure, instead of the more traditional analysis in which integrated intensities are used to derive experimental structure factors that are compared

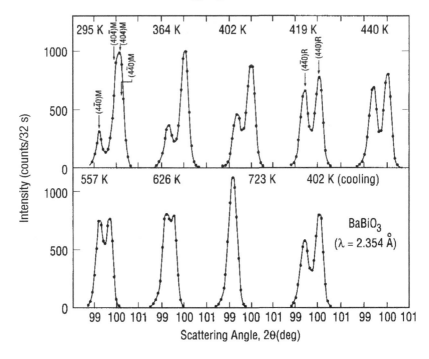

Fig. 7.21. Temperature dependence of the pseudocubic (440) reflection ($2a_0$ cell) of $BaBiO_3$ (from Cox and Sleight, 1979).

with calculated structure factors for an assumed model. The main advantage is at large Q where peaks inevitably tend to overlap. Specialized neutron powder diffraction instruments are located at every major neutron facility. Most employ multi-detecting schemes that are very efficient in collecting data and determining structures on line. While a triple-axis instrument cannot compete with these specialized counting techniques from the point of view of counting speed, it can provide advantages in terms of signal-to-noise ratio and variability of the momentum resolution. The three-axis instrument is especially useful for powder diffraction when only a limited angular region is being scanned, as, for example, in the study of a phase transition where new Bragg peaks appear.

D. Cox developed the use of a three-axis instrument for powder diffraction in the 1970s, and some details are given by Cox and Sleight (1979) in their study of $BaBiO_3$, a perovskite system that undergoes a series of structural phase transitions as a function of temperature. Evidence of these changes can be seen in Fig. 7.21, where the temperature-dependent splitting of the (440) Bragg peak is shown. Detection of the splitting required high angular resolution at $2\theta \sim 100°$. This was achieved by choosing the (004) reflection

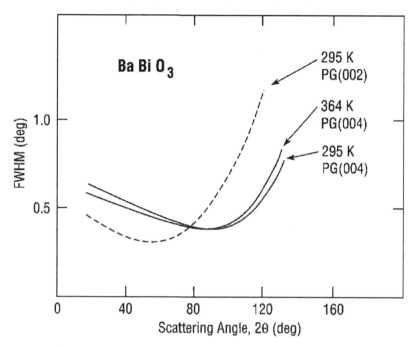

Fig. 7.22. Full-width at half-maximum of neutron powder peaks from BaBiO$_3$ as a function of scattering angle (2θ). Curves are calculated from the expression $H^2 = U \tan^2 \theta + V \tan \theta + W$, where U, V and W are refined values. Solid and broken lines refer to the spectrometer configurations described in text (from Cox and Sleight, 1979).

of the PG analyzer. The angular resolution as a function of scattering angle is compared for the PG (002) and (004) reflections in Fig. 7.22. With PG (002), the minimum width occurs at $2\theta \sim 60°$, whereas, by using PG (004), the minimum is shifted to higher angles, closer to the desired range. The resolution is enhanced by using tight collimation after the analyzer. The solid lines in Fig. 7.22 are for horizontal collimations of 20′-40′-40′-10′, which gives adequate intensity plus good resolution. The first and last collimators, between the reactor and monochromator, and analyzer and detector, respectively, are the ones that have the greatest effect on the resolution.

A systematic study of the variation in resolution for different analyzer choices was performed by Fincher (1979), and his results are given in Table 7.1. In this test, measurements were carried out on different single-crystal test samples with two different incident energies, $E_i = 14.7$ and $5 \, \text{meV}$; the same 10′-10′-10′-10′ configuration was used in all cases. The resolution is given in terms of the full-width at half-maximum (FWHM) of a longitudinal scan through a Bragg peak. The "Resolution at focusing position" given in

Table 7.1. Comparison of momentum resolutions obtainable with various analyzer choices. In all cases, monochromator is PG (002) and collimators are 10'-10'-Sample-10'-10'. Perfect Ge is used for B and E, and slightly deformed ($\eta = 2.3'$) Ge is used for F (after Fincher, 1979).

Setting	Analyzer	k_i (Å^{-1})	Resolution at focusing position (10^{-3} Å^{-1})	Sample	Reflection	2θ (°)	Measured resolution (10^{-3} Å^{-1})	Calculated resolution (10^{-3} Å^{-1})
A	PG (002)	2.67	6.96	none				
B	Ge (111)	2.67	0.22	none				
C	PG (004)	2.67	5.64	Nb$_3$Sn	(400)	125.7	6.18	6.72
D	PG (002)	1.55	3.35	V$_3$Si	(110)	74.8	2.80	3.54
E	Ge (111)	1.55	0.15	V$_3$Si	(110)	74.8	2.54	2.17
F	Ge (111)	1.55	1.40	K$_2$CuF$_4$	(004)	79.3	2.67	2.28

column 4 is that obtained at the angle at which the FWHM is calculated to be a minimum (see Fig. 7.22). Note that for problems where the scattering is diffuse in nature, such as critical scattering, it is the absolute value of ΔQ that is important, whereas in other cases, such as determination of lattice constants or resolution of a Bragg-peak splitting, it is $\Delta Q / Q$ that must be minimized.

Settings E and F compare the resolution obtainable with two different Ge (111) crystals: a perfect crystal in E and a crystal with a mosaic width of 0.04° in F. It is readily seen that the use of a perfect crystal can decrease the resolution width at the focusing condition by an order of magnitude. Of course, the gain in resolution comes at the cost of a loss in intensity of a similar magnitude. Settings A and B compare the calculated resolution at focusing position for PG (002) and Ge (111). Since the d-spacings for these two reflections are comparable, the difference in resolution is due entirely to differences in mosaic width. Setting C gives another example in which the use of PG (004) is advantageous. The last two columns of the table compare the measured and calculated resolutions, where the calculation was done with the Cooper–Nathans formalism outlined in Chap. 4. The agreement is generally within 20%.

References

Anderson, P. W. (1960). In *Fizika Dielectrikov*, ed. G. I. Skanavi (Acad. Science USSR, Moscow).

Axe, J. D. (1982). BNL Research Memo G-120 (unpublished).

Brockhouse, B. N. (1961). In *Inelastic Scattering of Neutrons in Solids and Liquids*, (International Atomic Energy Agency, Vienna) p. 113.

Chou, H., Schmidt, V. H., and Shapiro, S. M. (unpublished).

Cochran, W. (1960). *Advan. Phys.* **9**, 387.

Cowley, R. A., Buyers, W. J. L., and Dolling, G. (1969). *Solid State Commun.* **7**, 181.

Cox, D. E. and Sleight, A. W. (1979). *Acta Cryst.* B **35**, 1.

Dietrich, O. W., Cowley, R. A., and Shapiro, S. M. (1974). *J. Phys. C* **7**, L239.

Erickson, R. A. (1953). *Phys. Rev.* **90**, 779.

Fincher, Jr, C. R. (1979). BNL Research Memo G-110 (unpublished).

Goldman, A. I., Mohanty, K., Shirane, G., Horn, P. M., Greene, R. L., Peters, C. J., Thurston, T. R., and Birgeneau, R. J. (1987). *Phys. Rev. B* **36**, 5609.

Hamilton, W. C. (1958). *Phys. Rev.* **110**, 1050.

Hastings, J. B., and Batterman, B. W. (1975). *Phys. Rev. B* **12**, 5580.

Iizumi, M. (1973). *Jpn. J. Appl. Phys.* **12**, 167.

Iizumi, M. and Shirane, G. (1975). BNL Research Memo G-22 (unpublished).

Iizumi, M., Koetzle, T., Shirane, G., Chikazumi, S., Matsui, M., and Todo, S. (1982). *Acta Cryst. B* **38**, 2121.

Keating, D., Nunes, A., Batterman, B., and Hastings, J. (1971). *Phys. Rev. B* **4**, 2472.

Kjems, J. K. and Satija, S. K. (1981). BNL Research Memo G-20 Amendment (unpublished).

Meyer, G. M., Dietrich, O. W., Nelms, R. J., Hay, W. J., and Cowley, R. A. (1975). *J. Phys. C* **9**, L83.

Moncton, D., Lynn, J., and Shirane, G. (1974). BNL Research Memo G-20 (unpublished).

Moon, R. M. and Shull, C. G., (1964). *Acta Cryst.* **17**, 805.

Okazaki, A. and Willis, B. T. M. (1980). Harwell Report HL80/2640 (unpublished).

Pynn, R. (1975). *Acta Cryst. B* **31**, 2555.

Renninger, M. (1937). *Z. Phys.* **106**, 141.

Rietveld, H. M. (1969). *J. Appl. Cryst.* **2**, 65.

Shirane, G. and Yamada, Y. (1969). *Phys. Rev.* **177**, 858.

Shirane, G. and Axe, J. D. (1971). *Phys. Rev. B* **4**, 2957.

Shirane, G., Chikazumi, S., Akimitsu, J., Chiba, K., Matsui, M., and Fujii, Y. (1975). *J. Phys. Soc. Japan* **39**, 949.

Shirane, G., Fujii, Y., Hoshino, S., and Fujishita, H. (1981). (unpublished).

Tranquada, J. M., Moudden, A. H., Goldman, A. I., Zolliker, P., Cox, D. E., Shirane, G., Sinha, S. K., Vaknin, D., Johnston, D. C., Alvarez, M. S., Jacobson, A. J., Lewandowski, J. T., and Newsam, J. M. (1988). *Phys. Rev. B* **38**, 2477.

Trucano, P. and Batterman, B. W. (1972). *Phys. Rev. B* **6**, 3659.

Verwey, E. J. W. and Haayman, P. W. (1941). *Physica* **8**, 979.

Yamaguchi, Y., Kato, H., Takei, H., Goldman, A. I., and Shirane, G. (1986). *Solid State Commun.* **59**, 865.

8

Polarized neutrons

Among a variety of neutron scattering techniques, those utilizing polarized neutrons have shown rapid development over the last one to two decades. One of the most significant features of using polarized neutrons is the ability to separate magnetic cross sections from nuclear scattering processes. Thus, there exist many important experiments which can be performed only by such techniques. The use of a triple-axis spectrometer with polarization capabilities, which will be our main focus here, makes up a small, but important subset of polarized neutron experiments.

In Chap. 2 we discussed the magnetic scattering cross section for unpolarized neutrons. Here we start with a description of the additional information that can be gained with polarized neutrons. Of course, to take advantage of polarization analysis, one must have techniques for polarizing and manipulating polarized neutron beams, and we give a brief description of these. Turning to applications, we give a few examples of magnetic form-factor measurements in paramagnetic or ferromagnetic systems, which use techniques pioneered by Shull and his collaborators in the 1950s (Shull, Strauser, and Wollan, 1951). Finally, we discuss polarization analysis on a triple-axis spectrometer as developed by Moon, Riste, and Koehler (1969). The latter approach is quite powerful, as it allows one to uniquely isolate both the elastic and inelastic neutron cross sections; however, due to intensity limitations, polarization analysis was not fully utilized until the 1980s. A more extensive review of polarized neutron techniques, including neutron spin-echo spectroscopy, is given in the book *Polarized Neutrons* by Williams (1988).

8.1 Scattering of polarized neutrons

8.1.1 Definition of polarization

Let us begin by briefly summarizing the definition of polarization. The neutron is a spin-$\frac{1}{2}$ particle with two possible spin angular momentum values,

$\pm\frac{1}{2}\hbar$. Let the spin operator for a neutron be denoted as $\hat{\mathbf{s}} = \frac{1}{2}\boldsymbol{\sigma}$, where the components of $\boldsymbol{\sigma}$ are equal to the Pauli spin matrices. The polarization vector $\mathbf{P}$ of a neutron is equal to the expectation value of the spin operator divided by its maximum value

$$\mathbf{P} = \langle\hat{\mathbf{s}}\rangle / \left(\tfrac{1}{2}\right) = 2\langle\hat{\mathbf{s}}\rangle = \langle\boldsymbol{\sigma}\rangle. \tag{8.1}$$

In a neutron beam containing N spins, each with a polarization $\mathbf{P}_j$, the beam polarization is defined as follows,

$$\mathbf{P}_0 = \frac{1}{N}\left(\sum_j \mathbf{P}_j\right). \tag{8.2}$$

Since it is sufficient to consider the linear polarization vector in the direction of an applied magnetic field for most polarized neutron experiments, the beam polarization in the applied field direction, normally defined as the z-direction, can be expressed by

$$P_0 = \frac{N_+ - N_-}{N_+ + N_-} = n_+ - n_-, \tag{8.3}$$

where N_+ denotes number of "up" spins (defined as parallel to the field) and N_- the number of "down", or antiparallel, spins, with $N_+ + N_- = N$. It follows that n_+ and n_- are the probabilities of finding neutrons in the up and down spin states, respectively, with $n_+ + n_- = 1$. In terms of the beam polarization, we can write

$$n_+ = \tfrac{1}{2}(1 + P_0), \qquad n_- = \tfrac{1}{2}(1 - P_0). \tag{8.4}$$

8.1.2 Scattering formulas

We noted in §2.6, Eq. (2.61), that the differential cross section for scattering of polarized neutrons is given by

$$\left.\frac{d^2\sigma}{d\Omega_f dE_f}\right|_{s_i s_f} = \frac{k_f}{k_i}\sum_{i,f} P(i)\left|\langle f|\sum_l e^{i\mathbf{Q}\cdot\mathbf{r}_l}U_l^{s_i s_f}|i\rangle\right|^2 \delta(\hbar\omega + E_i - E_f), \tag{8.5}$$

with

$$U_l^{s_i s_f} = \langle s_f|b_l - p_l\mathbf{S}_{\perp l}\cdot\boldsymbol{\sigma} + B_l\mathbf{I}_l\cdot\boldsymbol{\sigma}|s_i\rangle. \tag{8.6}$$

Figure 8.1(a) provides a graphic definition of $\mathbf{S}_\perp$.

To avoid confusion with spectrometer coordinates, let us denote spatial coordinates with respect to the neutron polarization as (ξ, η, ζ), where we take the initial polarization to be along the ζ-axis. Moon, Riste, and Koehler

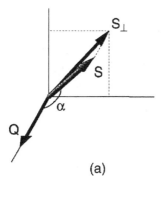

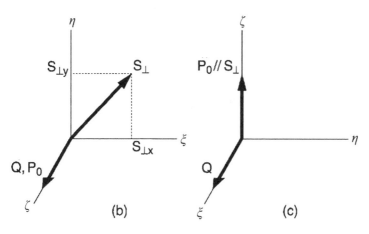

Fig. 8.1. Definition of the magnetic interaction vector $\mathbf{S}_\perp$, in relation to the scattering vector $\mathbf{Q}$ and the polarization vector $\mathbf{P}_0$ on the neutron.

(1969) pointed out that if one measures both the initial and final neutron polarizations with respect to the ζ-axis, then the four possible matrix elements $U^{\sigma\sigma}$ are given by

$$U^{++} = b - pS_{\perp\zeta} + BI_\zeta, \tag{8.7}$$
$$U^{--} = b + pS_{\perp\zeta} - BI_\zeta, \tag{8.8}$$
$$U^{+-} = -p\left(S_{\perp\xi} + iS_{\perp\eta}\right) + B\left(I_\xi + iI_\eta\right), \tag{8.9}$$
$$U^{-+} = -p\left(S_{\perp\xi} - iS_{\perp\eta}\right) + B\left(I_\xi - iI_\eta\right). \tag{8.10}$$

To simplify the discussion in the rest of this section, let us tentatively neglect the contribution from nuclear spins, $B\mathbf{I}$, until later in the chapter. Then, from the formulas for $U^{\sigma\sigma}$, we can identify two important cases. First, suppose that $\mathbf{Q}$ is parallel to $\mathbf{P}_0$, which is parallel to the ζ-axis. It follows that $S_{\perp\zeta} = 0$, and hence magnetic scattering appears in the *spin-flip* channels

(U^{+-} and U^{-+}). Such a case is illustrated in Fig. 8.1(b). Second, consider a condition in which both $\mathbf{Q} \perp \mathbf{P}_0$ [see Fig. 8.1(c)] and $\mathbf{S} \parallel \mathbf{P}_0$ are satisfied. In this case, $S_{\perp\xi} = S_{\perp\eta} = 0$, and hence all magnetic scattering occurs in the *non-spin-flip* channels (U^{++} and U^{--}). Note that the nuclear and magnetic scattering amplitudes contribute coherently; this is important for creating a polarized neutron beam and for measuring magnetic form factors.

For experiments in which the polarization is measured only along the ζ-axis, we can calculate the elastic non-spin-flip cross section $(d\sigma/d\Omega_f)^{\pm\pm}$ and the spin-flip cross section $(d\sigma/d\Omega)^{\pm\mp}$ by inserting Eqs. (8.7)–(8.10) into Eq. (8.5). We assume, for simplicity, perfect polarization for the incident neutron beam. Since $\mathbf{k}_i = \mathbf{k}_f$ and $E_i = E_f$ are satisfied for elastic scattering, we obtain

$$\left.\frac{d\sigma}{d\Omega_f}\right|_{\pm\pm} = \sum_{mn} e^{i\mathbf{Q}\cdot(\mathbf{r}_m-\mathbf{r}_n)}$$
$$\times \left[b_m b_n^* \pm \left(b_m p_n^* S_{n\perp\zeta}^* + b_n^* p_m S_{m\perp\zeta} + p_m p_n^* S_{m\perp\zeta} S_{n\perp\zeta}^*\right)\right] \quad (8.11)$$

$$\left.\frac{d\sigma}{d\Omega_f}\right|_{\pm\mp} = \sum_{mn} e^{i\mathbf{Q}\cdot(\mathbf{r}_m-\mathbf{r}_n)}$$
$$\times p_m p_n^* \left[S_{m\perp\xi} S_{n\perp\xi}^* + S_{m\perp\eta} S_{n\perp\eta}^* \mp i\zeta \cdot \left(\mathbf{S}_{m\perp} \times \mathbf{S}_{n\perp}^*\right)\right], \quad (8.12)$$

where the * indicates the complex conjugate. These equations verify that the ζ element of $\mathbf{S}_\perp$ contributes to the non-spin-flip scattering and the ξ and η elements to spin-flip scattering.

The total elastic differential cross section can be written using the incident polarization probabilities, n_+ and n_-, as

$$\frac{d\sigma}{d\Omega_f} = n_+ \left(\left.\frac{d\sigma}{d\Omega_f}\right|_{++} + \left.\frac{d\sigma}{d\Omega_f}\right|_{+-}\right) + n_- \left(\left.\frac{d\sigma}{d\Omega_f}\right|_{-+} + \left.\frac{d\sigma}{d\Omega}\right|_{--}\right). \quad (8.13)$$

In terms of the polarization vector, $\mathbf{P}_0$, this becomes

$$\frac{d\sigma}{d\Omega_f} = \sum_{mn} e^{i\mathbf{K}\cdot(\mathbf{r}_m-\mathbf{r}_n)} \left\{b_m b_n^* + p_m \mathbf{S}_{m\perp} \cdot p_n^* \mathbf{S}_{n\perp}^* \right.$$
$$\left. + \mathbf{P}_0 \cdot \left[b_m p_n^* \mathbf{S}_{n\perp}^* + b_n^* p_m \mathbf{S}_{m\perp} - i p_m p_n^* \left(\mathbf{S}_{m\perp} \times \mathbf{S}_{n\perp}^*\right)\right]\right\}. \quad (8.14)$$

The first and second terms in the summation correspond to the nuclear scattering and the magnetic scattering cross sections, respectively, for unpolarized neutrons ($\mathbf{P}_0 = 0$).

8.2 Elements of a polarized-beam spectrometer

8.2.1 *Polarizing a neutron beam*

Early studies of the magnetic scattering cross section were accomplished using the polarization-dependent transmission of a neutron beam passing through a magnetized iron slab. The total cross section σ_{scat} for scattering neutrons out of the beam depends on the relative orientation ($\pm$) of the neutron spin states with respect to the magnetization axis,

$$\sigma_{scat} = \sigma_0 \pm \sigma_P, \tag{8.15}$$

where σ_0 is the normal cross section of demagnetized iron, and σ_P is the polarization-dependent part. For a slab with atomic density N and thickness t, the transmitted intensity through the magnetized iron, I_{Mag}, is given by

$$\begin{aligned} I_{Mag} &= \tfrac{1}{2}I_0 e^{-Nt(\sigma_0+\sigma_P)} + \tfrac{1}{2}I_0 e^{-Nt(\sigma_0-\sigma_P)} \\ &= I_0 e^{-Nt\sigma_0}\cosh(Nt\sigma_P), \end{aligned} \tag{8.16}$$

while the normal (demagnetized) cross section I_{Demag} is simply

$$I_{Demag} = I_0 e^{-Nt\sigma_0}. \tag{8.17}$$

Thus the effect of the magnetic field is to produce an *increase* $\Delta I = I_{Mag} - I_{Demag}$ in the transmitted intensity,

$$\frac{\Delta I}{I_{Demag}} = \cosh(Nt\sigma_P) - 1 \approx \tfrac{1}{2}(Nt\sigma_P)^2, \tag{8.18}$$

where the approximation on the right assumes $Nt\sigma_P \ll 1$. Using Eq. (8.3), one can compute that the polarization of the transmitted beam is equal to $-\tanh(Nt\sigma_P) \approx Nt\sigma_P$. The polarization of the beam can be analyzed by transmission through a second magnetized iron crystal; this is called the *double* transmission effect.

It was also found that polarized neutrons can be produced using total external reflection from a magnetized mirror. The maximum angle for total reflection (the critical angle) is given by

$$\theta_c^{\pm} = \lambda(N/\pi)^{1/2}[\bar{b} \pm (B/B_s)\tilde{p}]^{1/2}, \tag{8.19}$$

where λ is the neutron wavelength, N is the atomic density of the mirror material, $\bar{b}$ is the average coherent scattering length, B and B_s are the applied and saturation fields, respectively, and

$$\tilde{p} = m_n \mu_N B_s / 2\pi\hbar^2 N. \tag{8.20}$$

For reflection angles in the range $\theta_c^- < \theta < \theta_c^+$ a fully polarized beam is obtained.

The more common technique of neutron polarization, utilizing a Bragg reflection from a magnetized monochromator, was developed by Shull, Strauser, and Wollan (1951). In order to discuss how a polarized beam is generated by a monochromator, let us consider a unit cell containing a single atom. Suppose we choose conditions such that $\mathbf{Q} \perp \mathbf{P}_0$ and $\mathbf{S} \parallel \mathbf{P}_0$ are satisfied. The latter condition requires

$$\mathbf{S} = \mathbf{S}_\perp = \left(0, 0, S_{\perp\zeta}\right). \tag{8.21}$$

Since $S_{\perp\xi}$ and $S_{\perp\eta}$ are zero, it is obvious from Eqs. (8.9) and (8.10) that $d\sigma/d\Omega|_{\pm\mp} = 0$; therefore, the total differential cross section becomes

$$\left.\frac{d\sigma}{d\Omega}\right|_{\pm\pm} = \left(b^2 \pm 2bpS + p^2S^2\right)$$

$$= \left(b \pm pS\right)^2. \tag{8.22}$$

For the general case of magnetized crystals, this can be written as

$$\left.\frac{d\sigma}{d\Omega}\right|_{\pm\pm} = \left(F_N \pm F_M\right)^2, \tag{8.23}$$

where F_N and F_M are the nuclear and magnetic structure factors, respectively. If one can find a special reflection with $b = pS$ or $F_N = F_M$, the diffracted neutron beam will be completely polarized since, for down $(-)$ spins, $b - p = 0$.

The Bragg reflection (220) from a magnetite (Fe_3O_4) single crystal was the first candidate for a polarizer tried by Nathans *et al.* (1959). Later, it was calculated that the (200) reflection of $Co_{0.92}Fe_{0.08}$ more nearly satisfies the condition $b = p$. A comparison of the experimentally-measured polarizing efficiencies for these two choices, as a function of applied field, is shown in Fig. 8.2. One drawback of Co–Fe is the absorption loss due to cobalt. The Co–alloy was later superseded by a single crystal of the Heusler alloy Cu_2MnAl (Delapalme *et al.*, 1971). The (111) reflection of the Heusler crystal has a much larger *d*-spacing, $d = 3.43\,\text{Å}$, compared with $1.76\,\text{Å}$ for the (200) reflection of $Co_{0.92}Fe_{0.08}$. Thus, the Heusler crystal offers a considerably better balance of intensity and resolution at low incident energies or longer neutron wavelengths. The successful growth of good single crystals of the Heusler alloy by Freund *et al.* (1983) at ILL is a key element of recent successes of polarization analysis techniques. For the polarized-beam instrument at ILL (IN20), five or eight rectangular crystals with dimensions $75 \times 15 \times 7\,\text{mm}^3$ are mounted with the longest axis horizontal on a flexible titanium backing plate, which is curved vertically (see § 3.2.1). The assembly of crystals is magnetized in a 0.12 T horizontal field supplied by $SmCo_5$

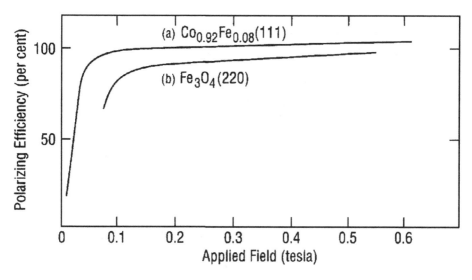

Fig. 8.2. Polarizing efficiencies of the magnetite (220) and $Co_{0.92}Fe_{0.08}$ (111) polarizing reflections as a function of applied field (from Nathans *et al.*, 1959).

permanent magnets. A very high reflectivity is obtained, as shown in Fig. 8.3. This vertically focused *reflection* geometry is an excellent setup for incident neutron energies less than 20 meV. At higher energies, the scattering angle becomes so small that the (111) reflection in *transmission* geometry is preferred.

Modern spin filters are also being used with increasing frequency. One type takes advantage of the fact that when a magnetic mirror is set to completely reflect up (+) spins, the neutrons passing through the mirror must be fully polarized in the opposite direction. By using a magnetic thin film on a silicon wafer, the mirror can be quite thin, allowing a high transmission, and clever design of multilayer films can significantly enhance θ_c. A transmission filter is created by using an array of closely spaced supermirrors, such as iron on silicon wafers (Majkrzak, 1995). Such a filter is especially useful for polarizing cold neutrons monochromatized with a high-reflectivity monochromator, such as pyrolytic graphite.

The latest development in transmission filters involves the use of spin-polarized ^3He. ^3He has a very strong absorption cross section for neutrons, which is why it is commonly used in neutron detectors; however, only a neutron with spin antiparallel to that of the neutron in a ^3He nucleus can be absorbed. Effective methods have been developed to prepare spin-polarized ^3He by optical pumping in strong magnetic fields; the polarized gas is transferred to a separate container for use on a spectrometer. The degree of

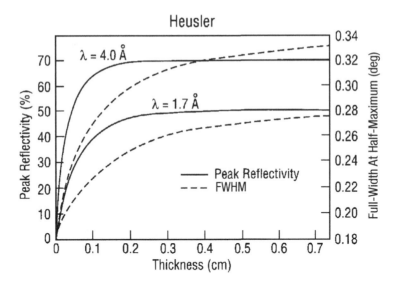

Fig. 8.3. Peak reflectivities (solid line) and full-width at half-maximum (FWHM, dashed line) of the peak profile as a function of crystal thickness at two different neutron wavelengths for a Heusler-alloy polarizing monochromator with a 0.186° mosaic width (from Freund *et al.*, 1983).

polarization and its lifetime are now sufficient that ^{3}He filters are in regular use at ILL (Heil *et al.*, 1999).

8.2.2 *Control of polarization direction*

Once the neutron beam is polarized, a small transverse magnetic field, called a guide field, is needed along the beam path to maintain the polarization. Homogeneity of the field is important to minimize depolarization. As will be discussed later, it is sometimes useful to be able to set the polarization direction at the sample to be either vertical or horizontal. This can be controlled with large-diameter pairs of solenoid coils (Helmholtz coils) centered on the sample axis. As long as the orientation of the guide field varies sufficiently slowly, the spin orientation will change adiabatically. The conditions for adiabatic rotation depend on the magnetic field and the velocity of the neutrons.

A rapid reversal of the magnetic field can cause the spin polarization to flip. For example, the alternation of magnetic field direction in the domains of a demagnetized ferromagnet can cause many spin flips and, when averaged over a sample, leads to depolarization of the neutron beam. Of course, it is generally useful to have devices which reverse the beam polarization in a

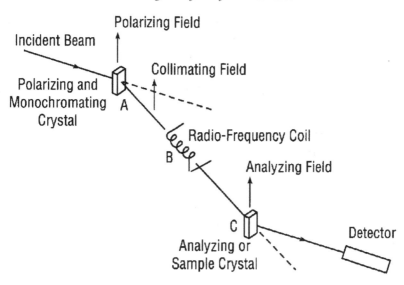

Fig. 8.4. A schematic diagram of a polarized-neutron spectrometer (from Nathans *et al.*, 1959).

controlled fashion, and a number of schemes have been employed in such spin-flippers.

Early work accomplished spin-flipping using a radio-frequency (rf) magnetic field; the strength of the field is tuned to adjust the adiabatic precession of the polarization to achieve the spin reversal. More popular today are spin-turn coils (frequently called Mezei flippers), which consist of flat solenoidal coils that create fields transverse to each guide field. The magnetic field of the flipper is adjusted so that neutrons of a selected wavelength will precess by exactly 180° while traversing the coil (see Pynn, 1984). Further alternatives are discussed by Williams (1988).

8.3 Magnetic form-factor studies

Just as one can use coherent interference between nuclear and magnetic scattering to generate a polarized beam by Bragg diffraction, the same effect can be used to study the magnetic scattering from a paramagnet or ferromagnet. The scheme can be implemented with a double-axis spectrometer; the original setup of Nathans *et al.* (1959) is shown in Fig. 8.4. The spectrometer consists of a polarizing monochromator, a flipper (the rf coil), and a guide (or collimating) field to prevent depolarization of the polarized beam by perturbing external fields. The sample is placed in a saturating magnetic field to align its magnetic moments.

In a typical experiment, one measures the intensity of a Bragg peak first with the flipper off and then with the flipper on. The ratio of these intensities, known as the flipping ratio, is given in the ideal case by

$$R = \left(\frac{F_N + F_M}{F_N - F_M}\right)^2.$$ (8.24)

In practice, there will be corrections due to incomplete polarization of the beam, imperfect flipper efficiency, and higher-order harmonics in the beam. Secondary extinction or depolarization from misoriented magnetic domains in the sample require further corrections. Corrections can be made in a straightforward way (see Williams, 1988).

The setup shown in Fig. 8.4 is often used to measure the polarization, P_0. For this task, the sample is replaced by a polarizing crystal such as Heusler alloy, in which case

$$R = \frac{n_+}{n_-} = \frac{1 + P_0}{1 - P_0},$$ (8.25)

assuming that the flipper performs ideally. With Heusler crystals one can attain $P_0 > 0.95$, which corresponds to $R > 39$. Note that R depends upon neutron energy, higher-order contamination, size of the beam, collimation, etc. It is not difficult to reach flipping ratios of $R > 100$ if the beam size is very small, e.g., $3\,\mathrm{mm}^2$; However, it is often difficult to obtain $R \sim 30$ when a beam size of $3\,\mathrm{cm}^2$ ($900\,\mathrm{mm}^2$) is needed. Nevertheless, a very high R is not essential for many polarized neutron measurements.

A particular strength of the flipping-ratio method is its sensitivity to weak magnetic scattering. When $F_M \ll F_N$, the flipping ratio reduces to

$$R \approx 1 + 4\frac{F_M}{F_N},$$ (8.26)

so that the flipping ratio is linear in F_M/F_N. In contrast, a measurement with an unpolarized beam yields a Bragg intensity that depends on $(F_M/F_N)^2$. A ratio of $F_M/F_N = 0.1$ would be extremely difficult to detect with unpolarized neutrons, whereas with polarized neutrons one would find $R = 1.4$.

The first extensive application of polarized neutron beams was the measurement of magnetic form factors $F_M(Q)$ pioneered by Shull, Strauser, and Wollan (1951). They first carried out measurements of the 3d ferromagnets Fe, Co, and Ni. Later, their studies were extended to several alloy crystals and rare-earth metals. As an example, the 3d form factor of Fe measured by Shull and Yamada (1962) is illustrated in Fig. 8.5; the results are shown only for the large-Q regime, where the form factor becomes quite small, and even negative. In a later study, Shull and Mook (1966) deduced that

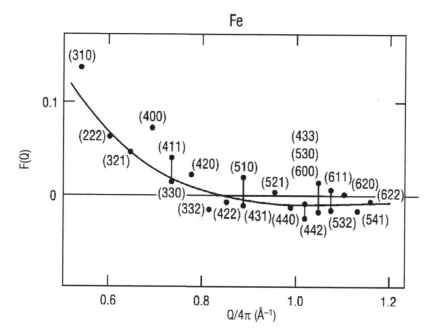

Fig. 8.5. Large-Q portion of the magnetic form factor of Fe measured by Shull and Yamada (1962).

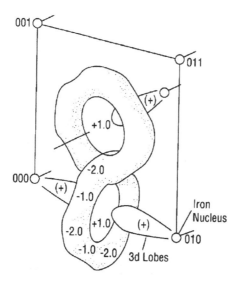

Fig. 8.6. Density distribution of negative magnetization deduced by Shull and Mook (1966) for the bcc unit cell of Fe. Positive 3d magnetization along edges of the cubic cell is exaggerated for comparison. Numbers denote the magnetization in the unit of 10^{-1} T.

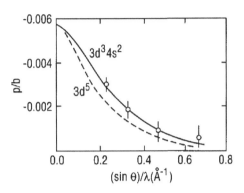

Fig. 8.7. Normalized magnetic form factor p/b of vanadium measured by Shull and Ferrier (1963). The magnetic scattering amplitude was measured as a function of the scattering angle θ under a magnetic field of up to 2.4 T, which aligns paramagnetic spins on V atoms.

narrow regions of negative magnetization occur in bcc Fe. The real-space distribution of the negative magnetization density is shown in Fig. 8.6.

The sensitivity of the flipping-ratio method is such that it is quite practical to measure the small moments induced in a paramagnet by a strong magnetic field. The magnetic scattering length, pS, for such a system can be written as

$$pS = 0.484\chi_s Hf(Q) \times 10^{-16} \text{ cm}, \qquad (8.27)$$

where $\chi_s H$ is the net magnetic moment per unit cell corresponding to the polarization of electronic spins induced by an applied field H, and $f(Q)$ is the magnetic form factor. Such a measurement was carried out by Shull and Ferrier (1963) on a vanadium single crystal. The results are shown in Fig. 8.7. The theoretical form factors of Watson and Freeman (1961) for $3d^5$ and $3d^3 4s^2$ configurations are also shown. Good agreement is obtained with the latter curve. The induced-moment technique still remains a very powerful tool to probe the nature of the magnetic electrons in solids. It was recently applied by Stassis *et al.* (1988) to the antiferromagnet La_2CuO_4, the parent material for the doped high-T_c superconductor $La_{1.85}Sr_{0.15}CuO_4$. Another example is the work of Lander *et al.* (1985) on the 5f electrons in Pu and U.

The flipping-ratio technique can also be implemented on a triple-axis spectrometer. Figure 8.8 shows a generic diagram for a triple-axis spectrometer, and the inset table lists four possible configurations. Alternative I, with a polarizing monochromator and no analyzer, corresponds to the double-axis case, discussed above. Equivalent configurations with three axes are given in II and III; in each of these, either the analyzer or the monochromator is a non-polarizing crystal such as pyrolytic graphite (PG). The loss

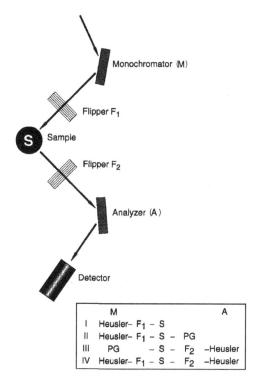

Fig. 8.8. Generic diagram for a polarized-neutron triple-axis spectrometer. Inset table lists several possible configurations covering two- and three-axis options. Guide fields are not indicated.

of intensity due to an extra crystal reflection is balanced by the improved signal-to-noise level. Conversion of a conventional triple-axis spectrometer to polarized setup III is particularly convenient, since the monochromator need not be changed (Shirane *et al.*, 1983), and one can take advantage of the focusing properties of a PG monochromator. Configuration IV is the orthodox triple-axis polarized-beam spectrometer with polarization analysis, as will be discussed in the next section.

8.4 Uniaxial polarization analysis

8.4.1 Basics of polarization analysis

Polarization analysis on a triple-axis spectrometer was advocated in the pioneering paper of Moon, Riste, and Koehler (1969). They constructed a triple-axis spectrometer with polarizing crystals on *both* the first and third axes at the Oak Ridge High Flux Isotope Reactor (HFIR) (see Fig. 8.9). In contrast to the double-axis instruments described in the previous section, which are restricted to measuring the magnetic contribution at Bragg peaks, the polarized-neutron triple-axis spectrometer enables one to measure the magnetic contribution to $S(\mathbf{Q}, \omega)$. Moon, Riste, and Koehler demonstrated the effectiveness of a simple one-dimensional polarization analysis for two particular conditions: $\mathbf{P}_0 \perp \mathbf{Q}$ and $\mathbf{P}_0 \parallel \mathbf{Q}$. The neutrons are polarized and analyzed along the vertical direction, and the orientation at the sample position can be controlled by a slow (adiabatic) rotation of the guide field. In Fig. 8.9, the electromagnet determines the polarization direction at the sample. [At a modern spectrometer, the electromagnet is replaced by orthogonal pairs of low-magnetic-field coils (Helmholtz coils).] It follows that the condition $\mathbf{P}_0 \perp \mathbf{Q}$ corresponds to a vertical field (VF) at the sample, while $\mathbf{P}_0 \parallel \mathbf{Q}$ is achieved with a horizontal field (HF).

As discussed in §8.1, there are four distinct matrix elements of the atomic scattering operator: U^{++}, U^{--}, U^{+-}, and U^{-+} [see Eqs. (8.7)–(8.10)]. The *non-spin-flip* matrix elements U^{++} and U^{--} are measured with both flippers off or both flippers on, respectively. The *spin-flip* matrix elements are measured when the first flipper is off and the second on, or vice versa, respectively.

It is obvious from the formulas for the matrix elements that the coherent nuclear scattering, i.e., Bragg and phonon scattering, is always non-spin-flip scattering. Since the isotopic incoherent scattering results from a randomness in the nuclear coherent scattering amplitude b, it is also non-spin-flip. As for magnetic and nuclear spins, the scattering is non-spin-flip for the effective spin components which are parallel to the neutron polarization direction $\mathbf{P}_0$, i.e., $S_{\perp\zeta}$ and I_ζ. The effective spin components perpendicular to $\mathbf{P}_0$, which are $S_{\perp\xi}, S_{\perp\eta}, I_\xi$, and I_η, are observed in spin-flip scattering. Taking into account that only those spin components perpendicular to the scattering vector are effective in neutron scattering, Moon, Riste, and Koehler (1969) observed that

If the neutron polarization $\mathbf{P}_0$ is along the scattering vector ($S_{\perp\zeta} = 0$), then all magnetic scattering is spin-flip scattering.

This statement is true for all types of scattering: *incoherent, coherent, elastic,* or *inelastic*. This offers us a simple tool for separating magnetic from non-magnetic scattering, which is one of the most important uses of polarized neutrons.

Figure 8.10 shows three examples of incoherent scattering measurements using polarized neutrons. In Fig. 8.10(a) the isotopic incoherent cross section of natural nickel is shown to appear only in the non-spin-flip channel; it can be completely eliminated by measuring with the spin-flip configuration. In

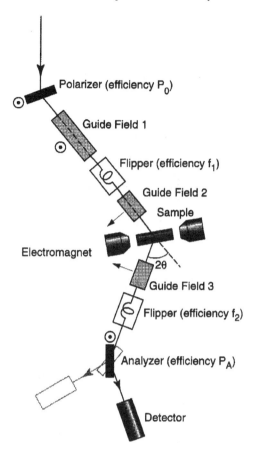

Fig. 8.9. Basic layout of a triple-axis spectrometer with polarization analysis capabilities (after Moon, Riste, and Koehler, 1969). For most applications, the electromagnet at the sample position is now replaced by a set of low-magnetic field coils which can switch the field direction between horizontal and vertical.

Fig. 8.10(b) the incoherent cross section of the unpolarized nuclear spins of vanadium is illustrated, where

$$\frac{d\sigma}{d\Omega}\bigg|_{++} = \frac{d\sigma}{d\Omega}\bigg|_{--} = \frac{1}{3}B^2 I(I+1), \qquad (8.28)$$

$$\frac{d\sigma}{d\Omega}\bigg|_{+-} = \frac{d\sigma}{d\Omega}\bigg|_{-+} = \frac{2}{3}B^2 I(I+1). \qquad (8.29)$$

In this case, the spin-flip scattering is twice as large as the non-spin-flip scattering, independent of the direction of the polarization. In Fig. 8.10(c) the paramagnetic scattering of MnF_2 is shown; further examples of paramagnetic

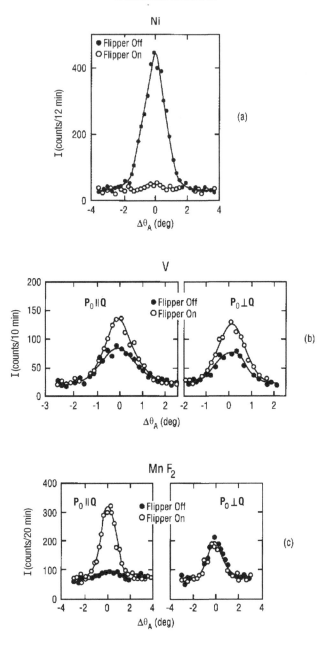

Fig. 8.10. Examples of polarization analysis, (after Moon, Riste, and Koehler 1969). The data were obtained by rocking the analyzer through the elastic position with fixed scattering angle. The "Flipper-Off" data are proportional to the $(++)$ cross section and the "Flipper-On" data are proportional to the $(-+)$ cross section. (a) Isotopic incoherent scattering from nickel. (b) Nuclear-spin incoherent scattering from vanadium. (c) Paramagnetic scattering from MnF_2.

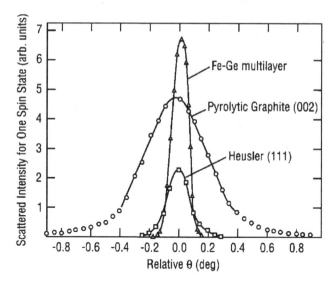

Fig. 8.11. Reflectivities and angular acceptances measured for an Fe–Ge multilayer, a pyrolytic graphite crystal (002) reflection, and a Heusler crystal (111) reflection at $k = 2.609$ Å^{-1} (from Majkrzak and Passell, 1985).

scattering are discussed in §8.4.3. The examples illustrated here suggest the wide range of applications of the polarization analysis technique.

The experimental arrangement shown in Fig. 8.9 is now well established and widely used. The most significant problem is the intensity loss in the polarization analysis arrangements, compared with the unpolarized triple-axis arrangement. A factor of 2 is lost because only only one of the two spin states is used. Approximately another factor of 2 is lost due to the poorer reflectivity of a Heusler monochromator instead of a PG monochromator. In addition, the focusing at the monochromator is more challenging with Heusler than with PG, and this could account for another factor of 5. Overall, the intensity ratio between the polarization analysis and the ordinary unpolarized triple-axis setup could reach 20–40. An intensity ratio of 8 has been attained at the IN20 instrument of ILL, where vertically focusing Heusler strips in a horizontal field is used for the monochromator. For cold neutrons, it is possible to take advantage of polarizing devices as supermirrors and multilayers. The reflectivity approaches 100% for a narrow angular range, as shown in Fig. 8.11.

An alternative configuration has been employed at the NG5 (SPINS) spectrometer at NCNR (Lee and Majkrzak, 1999). The polarizer consists of a PG monochromator followed by a supermirror transmission polarizer (Majkrzak, 1995); the polarization analysis is done with another supermirror transmission polarizer and no analyzer crystal. Because the transmission polarizer reflects one polarization while transmitting the other, both beams can be collected simultaneously with a position-sensitive detector.

8.4.2 Applications to Bragg diffraction

One of the simplest applications of the polarization analysis technique is to distinguish between magnetic Bragg reflections and nuclear peaks in an antiferromagnet. This possibility is based upon Eqs. (8.7)–(8.10), which state that the non-magnetic peaks will always involve non-spin-flip scattering, while the magnetic peaks will involve a mixture of spin-flip and non-spin-flip scattering. Let us consider the following two cases.

(A) $\mathbf{Q} \parallel \mathbf{P}_0$

In this case, the magnetic scattering will be spin-flip and the nuclear scattering will be non-spin-flip. From Eq. (8.11) and (8.12), the differential cross sections are (ignoring nuclear-spin scattering)

$$\left.\frac{d\sigma}{d\Omega}\right|_{\pm\pm} = \sum_{mn} e^{i\mathbf{K}\cdot(\mathbf{r}_m - \mathbf{r}_n)} b_m b_n^*, \tag{8.30}$$

$$\left.\frac{d\sigma}{d\Omega}\right|_{\pm\mp} = \sum_{mn} e^{i\mathbf{K}\cdot(\mathbf{r}_m - \mathbf{r}_n)}$$
$$\times p_m p_n^* \left[S_{m\perp\xi} S_{n\perp\xi}^* + S_{m\perp\eta} S_{n\perp\eta}^* \mp i\boldsymbol{\zeta} \cdot \left(\mathbf{S}_{m\perp} \times \mathbf{S}_{n\perp}^* \right) \right]. \tag{8.31}$$

This completely separates nuclear and magnetic peaks at a single temperature, in contrast to the usual unpolarized-beam approach, where the magnetic intensities are obtained by taking the difference between diffraction patterns measured below and above the Néel temperature T_N. The subtraction method can be rather ineffective when large Debye–Waller changes occur, or if the crystal structure changes below T_N.

An example of isothermal separation is shown in Fig. 8.12 for α-Fe_2O_3 powder. Note that the spin-flip cross section has lower background and that the aluminum peaks from the sample holder show up only in the non-spin-flip cross section. Although the subtraction method is not very difficult for this particular case of α-Fe_2O_3, there are some cases in which the separations are far more complicated. A good example is the antiferromagnetic order in V_2O_3 below the metal–insulator transition, which is accompanied by a structural transition. Recently, the spin-flip technique has been used more routinely in order to confirm the magnetic origin of weak peaks. For example, Mitsuda *et al.* (1987) established that the (100) reflection observed in La_2CuO_4 is indeed of magnetic origin.

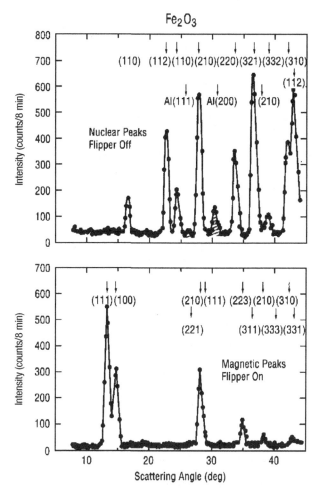

Fig. 8.12. α-Fe$_2$O$_3$ powder pattern. Separation of nuclear and magnetic peaks through polarization analysis at $\mathbf{Q} \parallel \mathbf{P}_0$ (from Moon, Riste, and Koehler, 1969).

(B) $\mathbf{Q} \perp \mathbf{P}_0$

Let us consider a coordinate system as illustrated in Fig. 8.1(c), where $S_{\perp\xi} = 0$. The differential cross sections are (neglecting nuclear-spin scattering)

$$\frac{d\sigma}{d\Omega}\bigg|_{\pm\pm} = \sum_{mn} e^{i\mathbf{K}\cdot(\mathbf{r}_m-\mathbf{r}_n)}$$
$$\times \left[b_m b_n^* \pm \left(b_m p_n^* S_{n\perp\zeta}^* + b_n^* p_m S_{m\perp\zeta} + p_m p_n^* S_{m\perp\zeta} S_{n\perp\zeta}^*\right)\right], \quad (8.32)$$

$$\frac{d\sigma}{d\Omega}\bigg|_{\pm\mp} = \sum_{mn} e^{i\mathbf{K}\cdot(\mathbf{r}_m-\mathbf{r}_n)}$$
$$\times p_m p_n^* S_{m\perp\eta} S_{n\perp\eta}^*. \quad (8.33)$$

Table 8.1. *Neutron cross sections for horizontal (HF) and vertical (VF)*
guide field at the sample position. Flipper ON and OFF correspond to
spin-flip (+−) and non-spin-flip (++) cross section, respectively. The
various contributions are: M = magnetic, NSI = nuclear spin incoherent,
N = nuclear (other than NSI), B = background.

Guide Field	ON (+−)	OFF (++)
HF	$M + \frac{2}{3}NSI + B$	$\frac{1}{3}NSI + N + B$
VF	$\frac{1}{2}M + \frac{2}{3}NSI + B$	$\frac{1}{2}M + \frac{1}{3}NSI + N + B$

In the non-spin-flip cross section, the atomic-spin components involved are
those perpendicular to the scattering vector $\mathbf{Q}$ *and* parallel to the nuclear
polarization $\mathbf{P_0}$. In contrast, the spin-flip cross section has the atomic-spin
components which are perpendicular to both $\mathbf{Q}$ and $\mathbf{P_0}$. Since the atomic
spins are aligned only along $\mathbf{P_0}$ in collinear ferromagnets, the spin-flip
scattering vanishes. Thus, this geometry is useful for searching for departures
from the collinearity. As summarized by Moon, Riste, and Koehler (1969),

One has only to remember that components of $\mathbf{S_\perp}$ *which are perpendicular to the*
neutron polarization will produce spin-flip scattering, while the component parallel to
the neutron polarization will produce non-spin-flip scattering.

8.4.3 *Paramagnetic scattering by the differential technique* (HF − VF)

One of the most important applications of polarization analysis is the study
of paramagnetic scattering. In many cases, it has turned out that this is
the only technique available to separate the magnetic cross section from all
the other unwanted scattering components. A modification of the setup in
Fig. 8.9 is used for the experiments. A guide field at the sample position
can be switched between the horizontal field (HF), i.e., $\mathbf{P_0} \parallel \mathbf{Q}$, and the
vertical field (VF), i.e., $\mathbf{P_0} \perp \mathbf{Q}$. Since only a small guide field of ~ 10 Oe
is required to maintain the neutron polarization, a pair of perpendicular
solenoid coils are usually used. Table 8.1 lists the four sets of cross sections
typically measured. ON and OFF indicate the state of the second flipper F2
when the final neutron energy E_f is fixed; ON corresponds to spin-flip (+−)
and OFF to non-spin-flip (++).

Figure 8.13 demonstrates the importance of polarization analysis for the
study of the paramagnetic scattering from iron (Wicksted, Böni, and Shirane,
1984). Figure 8.13(a) shows the unpolarized-neutron scattering below the

Curie temperature, $T_C = 1043\,\text{K}$, from an α-Fe (bcc) single crystal. The neutron beam is depolarized by domain walls below T_C, unless a strong saturating field is applied. As indicated in the figure, one can clearly see two peaks corresponding to the longitudinal acoustic (LA) phonon and magnon branches at $(1.12, 1.12, 0)$. Although the scan is intended to measure only the longitudinal modes, a weak transverse acoustic (TA) phonon branch is also seen because of a resolution effect (see §4.8). The significance of the polarized neutrons as well as the polarization analysis is clearly illustrated in Figs. 8.13(b) and (c). Figure 8.13(b) shows a constant-Q scan at $T_C + 22\,\text{K}\,(= 1.02\,T_C)$. The open circles represent the data obtained with the flipper ON and the horizontal field HF, which mainly consist of the paramagnetic scattering (see Table 8.1). Also shown is the nuclear scattering, represented by the open triangles, which were obtained with the flipper OFF. The separation between the nuclear and magnetic scattering is seen very clearly in this figure, though it should be noted that the "ON" data does contain additional contributions due to the nuclear incoherent scattering, the background, and some contaminations due to the finite flipping ratio. The open circles in Fig. 8.13(c), which were obtained with the flipper ON, result from the subtraction of data measured with the vertical field from those with the horizontal field, (HF $-$ VF). This difference (see Table 8.1) is purely magnetic and corresponds to one-half of the paramagnetic cross section. The differential technique (HF $-$ VF) is essential when the magnetic cross section becomes smaller than the instrumental background, e.g., the high energy-transfer data of Fig. 8.13(b). The counting time in constant-Q scans with E_f fixed changes considerably as a function of energy transfers, and background is generally proportional to counting time. These difficulties are completely eliminated in the differential data of Fig. 8.13(c). (The size of the crystal used for this experiment is $1.5\,\text{cm}^3$. The time taken to amass one million neutron counts (1 M monitor counts) corresponds to 1 minute at low energy transfers.)

The **Q** dependence of the paramagnetic scattering of Fe at $T = 1.02T_C$ is shown in Fig. 8.14. The dynamic scattering function can be written as a simple product of two lorentzians, one for the **Q** dependence and one for the energy dependence,

$$S(\mathbf{Q}, \omega) \sim \frac{1}{\kappa_1^2 + q^2} \frac{\Gamma}{\Gamma^2 + \omega^2} \frac{\omega/k_B T}{1 - e^{-\omega/k_B T}}, \tag{8.34}$$

$$\Gamma = Aq^{2.5} f(\kappa_1/q), \tag{8.35}$$

where $\mathbf{q} = \mathbf{Q} - \mathbf{G}$; the Résibois–Piette function $f(\kappa_1/q)$ is unity at T_C

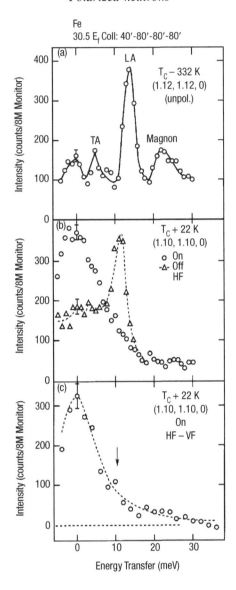

Fig. 8.13. (a) Unpolarized-neutron study of Fe at $\mathbf{Q} = (1.12, 1.12, 0)$ and $T \sim T_C - 332\,K$. Both flipper and external magnetic field are "OFF". (b) Polarized-beam study at $(1.10, 1.10, 0)$ and $T = 1.02T_C$ with horizontal magnetic field at the sample position. Open circles denote data obtained with "flipper-ON", whereas open triangles and dashed lines illustrate data obtained with "flipper-OFF". (c) Polarized-beam study by the differential technique (HF−VF) at $(1.10, 1.10, 0)$ and $T = 1.02T_C$ with "flipper-ON" (from Wicksted, Böni, and Shirane, 1984).

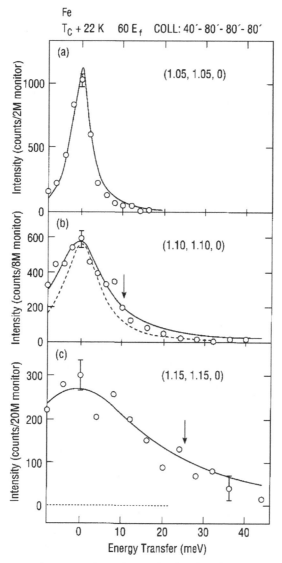

Fig. 8.14. Paramagnetic scattering data obtained using a differential technique (HF − VF) at $T = 1.02T_C$ at reciprocal-space positions (a) $(1.05, 1.05, 0)$, (b) $(1.10, 1.10, 0)$, and (c) $(1.15, 1.15, 0)$. The solid line in each figure denotes calculated values. The dashed line in (b) represents the normalized values without convolving an instrumental resolution function. Arrows in (b) and (c) point to expected positions for spin-wave peaks reported in constant-E scans (from Wicksted, Böni, and Shirane, 1984).

and becomes proportional to $\left(\kappa_1/q\right)^{1/2}$ in the hydrodynamic region. The polarization-analysis technique has enabled extensive studies of the paramagnetic scattering of 3d metals, and the qualitative applicability of Eqs. (8.34) and (8.35) has been thoroughly demonstrated.

To obtain a better quantitative description of the measurements, a somewhat generalized form has been considered (Wicksted, Böni, and Shirane, 1984), in which the lorentzian frequency dependence is replaced by a relaxation function $F(q, \omega)$, where

$$F(q, \omega) = \left(\frac{\Gamma}{\Gamma^2 + \omega^2} \right)^{\epsilon(\omega)}, \tag{8.36}$$

with

$$\Gamma = Aq^{\delta}. \tag{8.37}$$

Suppose we take $\epsilon(\omega) = 1$; then near T_C, where $f(\kappa_1/q) \approx 1$ and $\kappa_1 \approx 0$, the q dependence of $S(\mathbf{Q}, \omega)$ becomes

$$S(\mathbf{Q}, \omega) \sim \frac{q^{\delta-2}}{q^{2\delta} + (\omega/A)^2}. \tag{8.38}$$

This function has a peak at finite q as long as $\delta > 2$, and theoretically one expects $\delta = 5/2$. It follows that constant-E scans will show a peak at finite q, even though constant-$\mathbf{Q}$ scans are always peaked at $\omega = 0$.

The symbols in Fig. 8.15(a) show the positions of peaks measured in constant-E scans of Fe at T_C. The three dashed lines show calculated peak positions using parameters that give good fits to constant-$\mathbf{Q}$ scans and an exponent parameter $\delta = 2.1$–2.7. The calculated peak positions are not especially sensitive to the value of δ; however, the agreement with experiment is poor. To get better agreement, the empirical relation

$$\epsilon(\omega) = \begin{cases} 1, & |\omega| \leq \Gamma \\ 1 + \alpha[(|\omega| - \Gamma)/\Gamma], & |\omega| \geq \Gamma \end{cases} \tag{8.39}$$

was utilized. The solid line through the data points in Fig. 8.15(a) is a calculation with $\alpha = 0.1$ and $\delta = 2.5$; the agreement is quite good. Figure 8.15(b) shows calculated intensity contours for Fe at $T_C + 22\,\mathrm{K}$. Related studies of peak shapes on constant-E scans in paramagnetic nickel have been reported by Böni *et al.* (1993).

Another example of the use of polarization analysis is the study of the temperature-induced magnetism in FeSi. As shown at the top of Fig. 8.16, the magnetic susceptibility data indicate a remarkable temperature dependence, suggesting the presence of an energy gap. Several attempts to find the magnetic cross sections by unpolarized neutron beams were unsuccessful. Figure 8.17 shows the polarization-dependent data obtained by Tajima *et al.* (1988) on a large ($12\,\mathrm{cm}^3$) single crystal. It shows a typical enhancement of the magnetic intensity at $\Delta E = 0$ around the (011) reflection. These

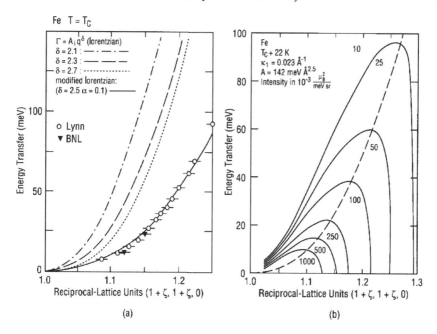

Fig. 8.15. (a) Dispersion of the peak positions measured in constant-E scans for Fe at T_C. Circles and triangles: experimental results; dashed lines: calculations using a lorentzian relaxation function with several values for the exponent parameter δ; solid line: calculation using the modified lorentzian of Eqs. (8.36) and (8.39) with $\alpha = 0.1$ and $\delta = 2.5$. (b) Intensity contours calculated with the modified lorentzian for $T = T_C + 22$ K. The dashed line through the contour maxima corresponds to the solid line in (a) (from Wicksted, Böni, and Shirane, 1984).

data were taken by the standard $(\mathrm{HF} - \mathrm{VF})$ differential technique. Note that the counting time is very long in spite of the huge volume of the crystal. In this particular case, only with the combination of the polarization analysis and the difference method can one properly extract the magnetic cross sections. In addition, the magnetic response spreads out over a very wide energy range, as shown in Fig. 8.18. The temperature dependence of the magnetic intensity is consistent with the susceptibility data (see lower panel of Fig. 8.16), confirming the conjecture of the temperature-induced magnetism.

There is one unique property in the magnetism of FeSi. Figure 8.17 shows a very strong $\mathbf{Q}$ dependence of the magnetic cross section $S(\mathbf{Q}, 0)$ around the nuclear peak at (011). The energy linewidth Γ is also strongly $\mathbf{Q}$ dependent as shown in Fig. 8.18(a). As a result, the energy-integrated cross section $S(\mathbf{Q}) = \int S(\mathbf{Q}, \omega) d\omega$ becomes $\mathbf{Q}$ independent, because the increase of Γ exactly cancels out the decrease of $S(q, 0)$. This indicates that there is no ferromagnetic correlation, contrary to what one would expect based on Fig. 8.17.

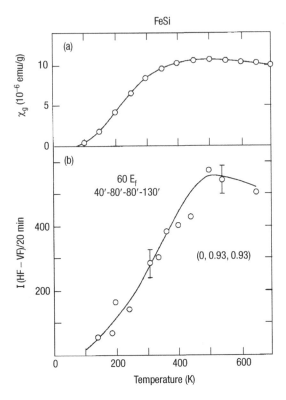

Fig. 8.16. Comparison, for FeSi, of (a) the magnetic susceptibility χ_g measured by Jaccarino *et al.* (1967) and (b) the magnetic scattering. The neutron data were taken near (011) with an energy resolution of 15 meV FWHM. The line is a guide to the eye (from Shirane *et al.*, 1987).

Another example of temperature-induced magnetism occurs in $LaCoO_3$, which has been studied by the same differential technique (Asai *et al.*, 1994). For Co^{3+}, with a $3d^6$ electronic configuration, it is possible for crystal-field effects to favor a low-spin ($S = 0$) state over the high-spin ($S = 2$) state normally expected according to Hund's rule. Conflicting interpretations of a low-spin to high-spin transition at either 90 K or 500 K existed in the literature. The data in Fig. 8.19 conclusively show that a moment is thermally induced on warming through 90 K. An analysis of thermal expansion and magnetic susceptibility data makes clear that the transition 90 K is to an intermediate-spin state, and that there is, in fact, a second spin-state transition near 500 K (Asai *et al.*, 1998).

Although we have so far quoted studies only from Brookhaven, there have, of course, been numerous polarization analysis experiments at other facilities, particularly ILL. For example, Kakurai *et al.* (1987) reported an

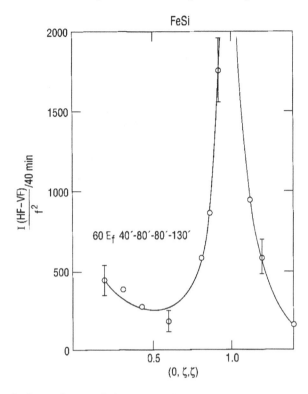

Fig. 8.17. The **Q** dependence of the magnetic scattering from FeSi along the [011] direction at 500 K, normalized to the square of the magnetic form factor for Fe. For reference, $d^*(011) = 1.98\,\text{Å}^{-1}$ at 300 K (from Shirane *et al.*, 1987).

interesting study of the one-dimensional ferromagnetic chain compound $CsNiF_3$ under an applied magnetic field. Boucher *et al.* (1990) carried out similar experiments on the antiferromagnetic chain compound TMMC. Pynn, Stirling, and Severing (1992) reported yet another field measurement on a single-Q sample of chromium in order to study the polarization of the 4 meV "commensurate" mode.

8.5 Spherical neutron polarimetry

We have seen that the uniaxial polarization analysis developed by Moon, Riste, and Koehler (1969) is extremely useful; however, it is not completely general. While it is frequently true that the polarization of scattered neutrons is parallel or antiparallel to the incident polarization, there are some cases in which the scattering can cause the polarization to be rotated to an arbitrary angle (Blume, 1963). In such situations, it is necessary to do a

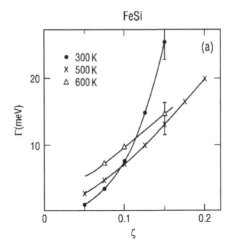

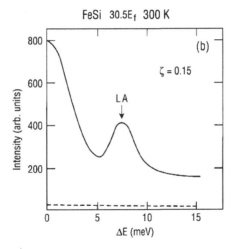

Fig. 8.18. (a) **Q** dependence of the energy linewidth Γ in FeSi at various temperatures, with $\mathbf{Q} = (1 + \zeta, 1 + \zeta, 0)$. Solid lines are the fits to $\Gamma = Aq^{\delta}$ with ($\bullet$) $A = 100 \, \text{meV} \, \text{Å}^3$, $\alpha = 3$ at 300 K, ($\times$) $A = 82 \, \text{meV} \, \text{Å}$, $\alpha = 1.5$ at 500 K, and ($\triangle$) $A = 50 \, \text{meV} \, \text{Å}$, $\alpha = 1$ at 600 K. (b) Comparison of the intensities of the nuclear (solid line) and magnetic (dashed line) scattering of $\zeta = 0.15$ at 300 K (from Tajima et al., 1988).

complete analysis of the final polarization vector, rather than measuring the projection along a single axis.

A zero-field polarimeter, called Cryopad, has been developed at ILL (Tasset *et al.*, 1999). Designed to be used on the conventional polarized triple-axis instrument IN20, it uses a Meissner shield to provide a zero-field region about the sample, with appropiate solenoid coils outside to rotate

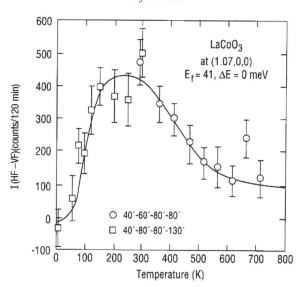

Fig. 8.19. Temperature dependence of the paramagnetic scattering cross section at $\mathbf{Q} = (1.07, 0, 0)$ with $E_f = 41$ meV, $\Delta E = 0$ meV for $LaCoO_3$. The solid line is a guide to the eye (from Asai *et al.*, 1994).

the incident and scattered polarizations. A second-generation polarimeter, Cryopad-II has a larger field-free zone, allowing the use of samples large enough for inelastic studies. Brown, Forsyth, and Tasset (1998) utilized Cryopad-II to study elastic diffraction from Cr_2O_3, which is a uniaxial antiferromagnet without a center of symmetry. They successully observed the expected polarization rotation; however, to do so they had to prepare the sample in a single magnetic domain by cooling in combined electric and magnetic fields, taking advantage of the magnetoelectric effect.

References

Asai, K., Yokokura, O., Nishimori, N., Chou, H., Tranquada, J. M., Shirane, G., Higuchi, S., Okajima, Y., and Kohn, K. (1994). *Phys. Rev. B* **50**, 3025.

Asai, K., Yoneda, A., Yokokura, O., Tranquada, J. M., Shirane, G., and Kohn, K. (1998). *J. Phys. Soc. Jpn.* **67**, 290.

Blume, M. (1963). *Phys. Rev.* **130**, 1670.

Böni, P., Mook, H. A., Martinez, J. L., and Shirane, G. (1993). *Phys. Rev. B* **47**, 3171.

Boucher, J. P., Pynn, R., Remoissenet, M., Regnault, L. P., Endoh, Y., and Renard, J. P. (1990). *Phys. Rev. Lett.* **64**, 1557.

Brown, P. J., Forsyth, J. B., and Tasset, F. (1998). *J. Phys.: Condens. Matter* **10**, 663 (1998).

Delapalme, A., Schweizer, J., Couderchon, G., and Perrier de la Bathie, R. (1971). *Nucl. Instrum. Methods* **95**, 589.

Freund, A., Pynn, R., Stirling, W. G., and Zeyen, C. M. E. (1983). *Physica B* **120**, 86.

Heil, W., Dreyer, J., Hofmann, D., Humblot, H., Lelièvre-Berna, E., and Tasset, F. (1999). *Physica B* **267–268**, 328.

Jaccarino, V., Wertheim, G. K., Wernick, J. H., Walker, L. R., and Arajs, S. (1967). *Phys. Rev.* **160**, 476.

Kakurai, K., Pynn, R., Steiner, M., and Dorner, B. (1987). *Phys. Rev. Lett.* **59**, 708.

Lander, G. H., Delapalme, A., Brown, P. J., Spinlet, J. C., Rebizant, J. C., and Vogt, O. (1985). *J. Appl. Phys.* **57**, 3748.

Lee, S.-H. and Majkrzak, C. F. (1999). *J. Neutron Research* **7**, 131.

Majkrzak, C. F. and Passell, L. (1985). *Acta Cryst. A* **41**, 41.

Majkrzak, C. F. (1995). *Physica B* **213–214**, 904.

Mitsuda, S., Shirane, G., Sinha, S. K., Johnston, D. C., Alvarez, M. S., Vaknin, D., and Moncton, D. E. (1987). *Phys. Rev. B* **36**, 822.

Moon, R. M., Riste, T., and Koehler, W. C. (1969). *Phys. Rev.* **181**, 920.

Nathans, R., Shull, C. G., Shirane, G., and Anderson, A. (1959). *J. Phys. Chem. Solids* **10**, 138.

Pynn, R. (1984). *Rev. Sci. Instrum.* **55**, 837.

Pynn, R., Stirling, W. G., and Severing, A. (1992). *Physica B* **180-181**, 203.

Shirane, G., Cowley, R. A., Majkrzak, C. F., Sokoloff, J. B., Pagonis, B. Perry, C. H., and Ishikawa, Y. (1983). *Phys. Rev. B* **28**, 6251.

Shirane, G., Fischer, J. E., Endoh, Y., and Tajima, K. (1987). *Phys. Rev. Lett.* **59**, 351.

Shull, C. G. and Ferrier, R. P. (1963). *Phys. Rev. Lett.* **10**, 295.

Shull, C. G. and Mook, H. F. (1966). *Phys. Rev. Lett.* **16**, 184.

Shull, C. G., Strauser, W. A., and Wollan, E. A. (1951). *Phys. Rev. B* **83**, 333.

Shull, C. G. and Yamada, Y. (1962). *J. Phys. Soc. Jpn* **17**, (BIII), 1.

Stassis, C., Harmon, B. N., Freltoft, T., Shirane, G., Sinha, S. K., Yamada, K., Endoh, Y., Hidaka, Y., and Murakami, T. (1988). *Phys. Rev. B* **38**, 9291.

Tajima, K., Endoh, Y., Fisher, J. E., and Shirane, G. (1988). *Phys. Rev. B* **38**, 6954.

Tasset, F., Brown, P. J., Lelièvre-Berna, E., Roberts, T., Pujol, S., Allibon, J., and Bourgeat-Lami, E. (1999). *Physica B* **267–268**, 69.

Watson, R. E. and Freeman, A. J. (1961). *Acta Cryst.* **14**, 27.

Wicksted, J. P., Böni, P., and Shirane, G. (1984). *Phys. Rev. B* **30**, 3655.

Williams, W. G. (1988). *Polarized Neutrons*, (Clarendon Press, Oxford).

Appendix 1

Neutron scattering lengths and cross sections

Table A1.1 lists neutron scattering lengths and cross sections taken from Sears (1992). Some of the notation has been introduced in Chap. 2, but a few notes of clarification are necessary.

The scattering length is, in general, complex, with the imaginary part corresponding to absorption. For an individual isotope, the coherent scattering length, b_{coh}, is equal to $\bar{b}$, the scattering length averaged over neutron and nuclear spin states, and the coherent scattering cross section is given by

$$\sigma_{coh} = 4\pi|b_{coh}|^2 = 4\pi|\bar{b}|^2. \tag{A1.1}$$

The total scattering cross section is given by

$$\sigma_{scat} = 4\pi\overline{|b|^2}, \tag{A1.2}$$

with the difference between σ_{scat} and σ_{coh} equal to the incoherent cross section. The incoherent scattering length is defined by

$$4\pi|b_{inc}|^2 = \sigma_{inc} = \sigma_{scat} - \sigma_{coh}. \tag{A1.3}$$

The spin incoherent scattering is identically zero for an isotope with no nuclear spin. Provided that the neutron and the nucleus are not both polarized, the absorption cross section is given by

$$\sigma_{abs} = \frac{4\pi}{k}b''_{coh}, \tag{A1.4}$$

where k is the magnitude of the neutron wave vector and b''_{coh} is the imaginary part of the coherent scattering length.

Because of the k-dependence of the absorption cross section (the so-called "$1/v$ law"), it is necessary to tabulate the cross section for some specific value of k, which is, by convention, $k = 3.494\,\text{Å}^{-1}$, corresponding to $\lambda = 1.798\,\text{Å}$ and $E = 25.30\,\text{meV}$. The scattering lengths are generally independent of k; the exceptions are ^{113}Cd, ^{149}Sm, ^{151}Eu, ^{155}Gd, ^{157}Gd, ^{176}Lu, and ^{180}Ta, each of which has a neutron-absorption, γ-emission resonance at thermal neutron energies. The scattering lengths and absorption cross sections in these latter cases are strongly energy dependent; the scattering lengths of the resonant rare-earth nuclides are tabulated as a function of energy in Lynn and Seeger (1990).

For an element with more than one isotope, the effective scattering lengths and cross sections involve an average over the natural distribution of isotopes. Let c_r be the relative natural abundance of the rth isotope, such that

$$\sum_r c_r = 1; \tag{A1.5}$$

then

$$b_{coh} = \sum_r c_r b_{coh,r}, \tag{A1.6}$$

$$\sigma_{scat} = \sum_r c_r \sigma_{scat,r}, \tag{A1.7}$$

$$\sigma_{abs} = \sum_r c_r \sigma_{abs,r}. \tag{A1.8}$$

The coherent and incoherent cross sections are still defined by Eqs. (A1.1) and (A1.3), respectively. It follows that

$$\sigma_{inc} = \sigma_{inc}(\text{spin}) + \sigma_{inc}(\text{isotope}), \tag{A1.9}$$

where

$$\sigma_{inc}(\text{spin}) = \sum_r c_r \sigma_{inc,r} = 4\pi \sum_r c_r |b_{inc,r}|^2 \tag{A1.10}$$

and

$$\sigma_{inc}(\text{isotope}) = 4\pi \sum_{r<r'} c_r c_{r'} |b_{coh,r} - b_{coh,r'}|^2. \tag{A1.11}$$

Note that for a mixture of isotopes, Eq. (A1.9) defines only the magnitude of b_{inc} and not its sign.

In Table A1.1, the scattering lengths are listed in units of femtometers, fm ($= 10^{-15}$ m $= 10^{-13}$ cm), and the cross sections in barns, where 1 barn $= 100$ fm^2 $= 10^{-24}$ cm^2. The values in parentheses are estimated standard errors. In a few cases of isotopes of low abundance where measurements were absent, scattering lengths are estimates, and such values are indicated by the letter "E".

References

Sears, V. F. (1992). *Neutron News* **3** (3), 303.
Lynn, J. E. and Seeger, P. A. (1990). *Atomic and Nuclear Data Tables* **44**, 191.

Table A1.1. *Neutron scattering lengths and cross sections. For unstable isotopes, the lifetime is listed in parentheses under "Natural abundance"* (a = annum).

Isotope	Natural abundance (%)	b_{coh} (fm)	b_{inc} (fm)	σ_{coh} (barn)	σ_{inc} (barn)	σ_{scat} (barn)	σ_{abs} (barn)
H		−3.7390(11)		1.7568(10)	80.26(6)	82.02(6)	0.3326(7)
^{1}H	99.985	−3.7406(11)	25.274(9)	1.7583(10)	80.27(6)	82.03(6)	0.3326(7)
^{2}H	0.015	6.671(4)	4.04(3)	5.592(7)	2.05(3)	7.64(3)	0.000519(7)
^{3}H	(12.32 a)	4.792(27)	−1.04(17)	2.89(3)	0.14(4)	3.03(5)	0
He		3.26(3)		1.34(2)	0	1.34(2)	0.00747(1)
^{3}He	0.00014	5.74(7) − 1.483(2)i	−2.5(6) + 2.568(3)i	4.42(10)	1.6(4)	6.0(4)	5333.(7.)
^{4}He	99.99986	3.26(3)	0	1.34(2)	0	1.34(2)	0
Li		−1.90(2)		0.454(10)	0.92(3)	1.37(3)	70.5(3)
^{6}Li	7.5	2.00(11) − 0.261(1)i	−1.89(10) + 0.26(1)i	0.51(5)	0.46(5)	0.97(7)	940.(4.)
^{7}Li	92.5	−2.22(2)	−2.49(5)	0.619(11)	0.78(3)	1.40(3)	0.0454(3)
^{9}Be	100	7.79(1)	0.12(3)	7.63(2)	0.0018(9)	7.63(2)	0.0076(8)
B		5.30(4) − 0.213(2)i		3.54(5)	1.70(12)	5.24(11)	767.(8.)
^{10}B	20.0	−0.1(3) − 1.066(3)i	−4.7(3) + 1.231(3)i	0.144(8)	3.0(4)	3.1(4)	3835.(9.)
^{11}B	80.0	6.65(4)	−1.3(2)	5.56(7)	0.21(7)	5.77(10)	0.0055(33)
C		6.6460(12)		5.550(2)	0.001(4)	5.551(3)	0.00350(7)
^{12}C	98.90	6.6511(16)	0	5.559(3)	0	5.559(3)	0.00353(7)
^{13}C	1.10	6.19(9)	−0.52(9)	4.81(14)	0.034(11)	4.84(14)	0.00137(4)
N		9.36(2)		11.01(5)	0.5(12)	11.51(11)	1.90(3)
^{14}N	99.63	9.37(2)	2.0(2)	11.03(5)	0.5(1)	11.53(11)	1.91(3)
^{15}N	0.37	6.44(3)	−0.02(2)	5.21(5)	0.00005(10)	5.21(5)	0.0000248(8)
O		5.803(4)		4.232(6)	0.000(8)	4.232(6)	0.00019(2)
^{16}O	99.762	5.803(4)	0	4.232(6)	0	4.232(6)	0.00010(2)
^{17}O	0.038	5.78(15)	0.18(6)	4.20(22)	0.004(3)	4.20(22)	0.236(10)
^{18}O	0.200	5.84(7)	0	4.29(10)	0	4.29(10)	0.00016(1)

Table A1.1. (cont.)

Isotope	Natural abundance (%)	b_{coh} (fm)	b_{inc} (fm)	σ_{coh} (barn)	σ_{inc} (barn)	σ_{scat} (barn)	σ_{abs} (barn)
^{19}F	100	5.654(10)	−0.082(9)	4.017(14)	0.0008(2)	4.018(14)	0.0096(5)
Ne		4.566(6)		2.620(7)	0.008(9)	2.628(6)	0.039(4)
^{20}Ne	90.51	4.631(6)	0	2.695(7)	0	2.695(7)	0.036(4)
^{21}Ne	0.27	6.66(19)	±0.6(1)	5.6(3)	0.05(2)	5.7(3)	0.67(11)
^{22}Ne	9.22	3.87(1)	0	1.88(1)	0	1.88(1)	0.046(6)
^{23}Na	100	3.63(2)	3.59(3)	1.66(2)	1.62(3)	3.28(4)	0.530(5)
Mg		5.375(4)		3.631(5)	0.08(6)	3.71(4)	0.063(3)
^{24}Mg	78.99	5.66(3)	0	4.03(4)	0	4.03(4)	0.050(5)
^{25}Mg	10.00	3.62(14)	1.48(10)	1.65(13)	0.28(4)	1.93(14)	0.19(3)
^{26}Mg	11.01	4.89(15)	0	3.00(18)	0	3.00(18)	0.0382(8)
^{27}Al	100	3.449(5)	0.256(10)	1.495(4)	0.0082(6)	1.503(4)	0.231(3)
Si		4.1491(10)		2.1633(10)	0.004(8)	2.167(8)	0.171(3)
^{28}Si	92.23	4.107(6)	0	2.120(6)	0	2.120(6)	0.177(3)
^{29}Si	4.67	4.70(10)	0.09(9)	2.78(12)	0.001(2)	2.78(12)	0.101(14)
^{30}Si	3.10	4.58(8)	0	2.64(9)	0	2.64(9)	0.107(2)
^{31}P	100	5.13(1)	0.2(2)	3.307(13)	0.005(10)	3.312(16)	0.172(6)
S		2.847(1)		1.0186(7)	0.007(5)	1.026(5)	0.53(1)
^{32}S	95.02	2.804(2)	0	0.9880(14)	0	0.9880(14)	0.54(4)
^{33}S	0.75	4.74(19)	1.5(1.5)	2.8(2)	0.3(6)	3.1(6)	0.54(4)
^{34}S	4.21	3.48(3)	0	1.52(3)	0	1.52(3)	0.227(5)
^{36}S	0.02	3.(1.)E	0	1.1(8)	0	1.1(8)	0.15(3)
Cl		9.5770(8)		11.526(2)	5.3(5)	16.8(5)	33.5(3)
^{35}Cl	75.77	11.65(2)	6.1(4)	17.06(6)	4.7(6)	21.8(6)	44.1(4)
^{37}Cl	24.23	3.08(6)	0.1(1)	1.19(5)	0.001(3)	1.19(5)	0.433(6)

Ar		1.909(6)		0.458(3)	0.225(5)	0.683(4)	0.675(9)
36Ar	0.337	24.90(7)	0	77.9(4)	0	77.9(4)	5.2(5)
38Ar	0.063	3.5(3.5)	0	1.5(3.1)	0	1.5(3.1)	0.8(2)
40Ar	99.600	1.830(6)	0	0.421(3)	0	0.421(3)	0.660(9)
K		3.67(2)		1.69(2)	0.27(11)	1.96(11)	2.1(1)
39K	93.258	3.74(2)	1.4(3)	1.76(2)	0.25(11)	2.01(11)	2.1(1)
40K	0.012	3.(1.)E		1.1(8)	0.5(5)E	1.6(9)	35.(8.)
41K	6.730	2.69(8)	1.5(1.5)	0.91(5)	0.3(6)	1.2(6)	1.46(3)
Ca		4.70(2)		2.78(2)	0.05(3)	2.83(2)	0.43(2)
40Ca	96.941	4.80(2)	0	2.90(2)	0	2.90(2)	0.41(2)
42Ca	0.647	3.36(10)	0	1.42(8)	0	1.42(8)	0.68(7)
43Ca	0.135	-1.56(9)		0.31(4)	0.5(5)E	0.8(5)	6.2(6)
44Ca	2.086	1.42(6)	0	0.25(2)	0	0.25(2)	0.88(5)
46Ca	0.004	3.6(2)	0	1.6(2)	0	1.6(2)	0.74(7)
48Ca	0.187	0.39(9)	0	0.019(9)	0	0.019(9)	1.09(14)
45Sc	100	12.29(11)	-6.0(3)	19.0(3)	4.5(5)	23.5(6)	27.5(2)
Ti		-3.438(2)		1.485(2)	2.87(3)	4.35(3)	6.09(13)
46Ti	8.2	4.93(6)	0	3.05(7)	0	3.05(7)	0.59(18)
47Ti	7.4	3.63(12)	-3.5(2)	1.66(11)	1.5(2)	3.2(2)	1.7(2)
48Ti	73.8	-6.08(2)	0	4.65(3)	0	4.65(3)	7.84(25)
49Ti	5.4	1.04(5)	5.1(2)	0.14(1)	3.3(3)	3.4(3)	2.2(3)
50Ti	5.2	6.18(8)	0	4.80(12)	0	4.80(12)	0.179(3)
V		-0.3824(12)		0.01838(12)	5.08(6)	5.10(6)	5.08(4)
50V	0.250	7.6(6)		7.3(1.1)	0.5(5)E	7.8(1.0)	60.(40.)
51V	99.750	-0.402(2)	6.35(4)	0.0203(2)	5.07(6)	5.09(6)	4.9(1)

Table A1.1. (*cont.*)

Isotope	Natural abundance (%)	b_{coh} (fm)	b_{inc} (fm)	σ_{coh} (barn)	σ_{inc} (barn)	σ_{scat} (barn)	σ_{abs} (barn)
Cr		3.635(7)		1.660(6)	1.83(2)	3.49(2)	3.05(8)
^{50}Cr	4.35	−4.50(5)	0	2.54(6)	0	2.54(6)	15.8(2)
^{52}Cr	83.79	4.920(10)	0	3.042(12)	0	3.042(12)	0.76(6)
^{53}Cr	9.50	−4.20(3)	6.87(10)	2.22(3)	5.93(17)	8.15(17)	18.1(1.5)
^{54}Cr	2.36	4.55(10)	0	2.60(11)	0	2.60(11)	0.36(4)
^{55}Mn	100	−3.73(2)	1.79(4)	1.75(2)	0.40(2)	2.15(3)	13.3(2)
Fe		9.45(2)		11.22(5)	0.40(11)	11.62(10)	2.56(3)
^{54}Fe	5.8	4.2(1)	0	2.2(1)	0	2.2(1)	2.25(18)
^{56}Fe	91.7	9.94(1)	0	12.42(7)	0	12.42(7)	2.59(14)
^{57}Fe	2.2	2.3(1)		0.66(6)	0.3(3)E	1.0(3)	2.48(30)
^{58}Fe	0.3	15.(7.)	0	28.(26.)	0	28.(26.)	1.28(5)
^{59}Co	100	2.49(2)	−6.2(2)	0.779(13)	4.8(3)	5.6(3)	37.18(6)
Ni		10.3(1)		13.3(3)	5.2(4)	18.5(3)	4.49(16)
^{58}Ni	68.27	14.4(1)	0	26.1(4)	0	26.1(4)	4.6(3)
^{60}Ni	26.10	2.8(1)	0	0.99(7)	0	0.99(7)	2.9(2)
^{61}Ni	1.13	7.60(6)	±3.9(3)	7.26(11)	1.9(3)	9.2(3)	2.5(8)
^{62}Ni	3.59	−8.7(2)	0	9.5(4)	0	9.5(4)	14.5(3)
^{64}Ni	0.91	−0.37(7)	0	0.017(7)	0	0.017(7)	1.52(3)
Cu		7.718(4)		7.485(8)	0.55(3)	8.03(3)	3.78(2)
^{63}Cu	69.17	6.43(15)	0.22(2)	5.2(2)	0.006(1)	5.2(2)	4.50(2)
^{65}Cu	30.83	10.61(19)	1.79(10)	14.1(5)	0.40(4)	14.5(5)	2.17(3)
Zn		5.680(5)		4.054(7)	0.077(7)	4.131(10)	1.11(2)
^{64}Zn	48.6	5.22(4)	0	3.42(5)	0	3.42(5)	0.93(9)
^{66}Zn	27.9	5.97(5)	0	4.48(8)	0	4.48(8)	0.62(6)
^{67}Zn	4.1	7.56(8)	−1.50(7)	7.18(15)	0.28(3)	7.46(15)	6.8(8)

^{68}Zn	18.8	6.03(3)	0	4.57(5)	0	4.57(5)	1.1(1)
^{70}Zn	0.6	6.(1.)E	0	4.5(1.5)	0	4.5(1.5)	0.092(5)
Ga		7.288(2)		6.675(4)	0.16(3)	6.83(3)	2.75(3)
^{69}Ga	60.1	7.88(2)	−0.85(5)	7.80(4)	0.091(11)	7.89(4)	2.18(5)
^{71}Ga	39.9	6.40(3)	−0.82(4)	5.15(5)	0.084(8)	5.23(5)	3.61(10)
Ge		8.185(20)		8.424(4)	0.18(7)	8.60(6)	2.20(4)
^{70}Ge	20.5	10.0(1)	0	12.6(3)	0	12.6(3)	3.0(2)
^{72}Ge	27.4	8.51(10)	0	9.1(2)	0	9.1(2)	0.8(2)
^{73}Ge	7.8	5.02(4)	3.4(3)	3.17(5)	1.5(3)	4.7(3)	15.1(4)
^{74}Ge	36.5	7.58(10)	0	7.2(2)	0	7.2(2)	0.4(2)
^{76}Ge	7.8	8.2(1.5)	0	8.(3.)	0	8.(3.)	0.16(2)
^{75}As	100	6.58(1)	−0.69(6)	5.44(2)	0.060(10)	5.50(2)	4.5(1)
Se		7.970(9)		7.98(2)	0.32(6)	8.30(6)	11.7(2)
^{74}Se	0.9	0.8(3.0)	0	0.1(6)	0	0.1(6)	51.8(1.2)
^{76}Se	9.0	12.2(1)	0	18.7(3)	0	18.7(3)	85.(7.)
^{77}Se	7.6	8.25(8)	±0.6(1.6)	8.6(2)	0.05(26)	8.65(16)	42.(4.)
^{78}Se	23.5	8.24(9)	0	8.5(2)	0	8.5(2)	0.43(2)
^{80}Se	49.6	7.48(3)	0	7.03(6)	0	7.03(6)	0.61(5)
^{82}Se	9.4	6.34(8)	0	5.05(13)	0	5.05(13)	0.044(3)
Br		6.795(15)		5.80(3)	0.10(9)	5.90(9)	6.9(2)
^{79}Br	50.69	6.80(7)	−1.1(2)	5.81(12)	0.15(6)	5.96(13)	11.0(7)
^{81}Br	49.31	6.79(7)	0.6(1)	5.79(12)	0.05(2)	5.84(12)	2.7(2)
Kr		7.81(2)		7.67(4)	0.01(14)	7.68(13)	25.(1.)
^{78}Kr	0.35		0		0		6.4(9)
^{80}Kr	2.25		0		0		11.8(5)

Table A1.1. (cont.)

Isotope	Natural abundance (%)	b_{coh} (fm)	b_{inc} (fm)	σ_{coh} (barn)	σ_{inc} (barn)	σ_{scat} (barn)	σ_{abs} (barn)
82Kr	11.6		0		0		29.(20.)
83Kr	11.5						185.(30.)
84Kr	57.0		0		0		0.113(15)
86Kr	17.3	8.1(2)	0	8.2(4)	0	8.2(4)	0.003(2)
Rb		7.09(2)		6.32(4)	0.5(4)	6.8(4)	0.38(1)
85Rb	72.17	7.03(10)		6.2(2)	0.5(5)E	6.7(5)	0.48(1)
87Rb	27.83	7.23(12)		6.6(2)	0.5(5)E	7.1(5)	0.12(3)
Sr		7.02(2)		6.19(4)	0.06(11)	6.25(10)	1.28(6)
84Sr	0.56	7.(1.)E	0	6.(2.)	0	6.(2.)	0.87(7)
86Sr	9.86	5.67(5)	0	4.04(7)	0	4.04(7)	1.04(7)
87Sr	7.00	7.40(7)		6.88(13)	0.5(5)E	7.4(5)	16.(3.)
88Sr	82.58	7.15(6)	0	6.42(11)	0	6.42(11)	0.058(4)
89Y	100	7.75(2)	1.1(3)	7.55(4)	0.15(8)	7.70(9)	1.28(2)
Zr		7.16(3)		6.44(5)	0.02(15)	6.46(14)	0.185(3)
90Zr	51.45	6.4(1)	0	5.1(2)	0	5.1(2)	0.011(5)
91Zr	11.32	8.7(1)	−1.08(15)	9.5(2)	0.15(4)	9.7(2)	1.17(10)
92Zr	17.19	7.4(2)	0	6.9(4)	0	6.9(4)	0.22(6)
94Zr	17.28	8.2(2)	0	8.4(4)	0	8.4(4)	0.0499(24)
96Zr	2.76	5.5(1)	0	3.8(1)	0	3.8(1)	0.0229(10)
93Nb	100	7.054(3)	−0.139(10)	6.253(5)	0.0024(3)	6.255(5)	1.15(5)
Mo		6.715(20)		5.67(3)	0.04(5)	5.71(4)	2.48(4)
92Mo	14.84	6.91(8)	0	6.00(14)	0	6.00(14)	0.019(2)
94Mo	9.25	6.80(7)	0	5.81(12)	0	5.81(12)	0.015(2)
95Mo	15.92	6.91(6)		6.00(10)	0.5(5)E	6.5(5)	13.1(3)
96Mo	16.68	6.20(6)	0	4.83(9)	0	4.83(9)	0.5(2)

Isotope	Conc.	b_c	b_i	σ_{coh}	σ_{inc}	σ_s	σ_a
^{97}Mo	9.55	7.24(8)		6.59(15)	0.5(5)E	7.1(5)	2.5(2)
^{98}Mo	24.13	6.58(7)	0	5.44(12)	0	5.44(12)	0.127(6)
^{100}Mo	9.63	6.73(7)	0	5.69(12)	0	5.69(12)	0.4(2)
^{99}Tc	(2.13 × 10^5 a)	6.8(3)		5.8(5)	0.5(5)E	6.3(7)	20.(1.)
Ru		7.03(3)		6.21(5)	0.4(1)	6.6(1)	2.56(13)
^{96}Ru	5.5		0		0		0.28(2)
^{98}Ru	1.9		0		0		< 8
^{99}Ru	12.7						6.9(1.0)
^{100}Ru	12.6		0		0		4.8(6)
^{101}Ru	17.0						3.3(9)
^{102}Ru	31.6		0		0		1.17(7)
^{104}Ru	18.7		0		0		0.31(2)
^{103}Rh	100	5.88(4)		4.34(6)	0.3(3)E	4.6(3)	144.8(7)
Pd		5.91(6)		4.39(9)	0.093(9)	4.48(9)	6.9(4)
^{102}Pd	1.02	7.7(7)E	0	7.5(1.4)	0	7.5(1.4)	3.4(3)
^{104}Pd	11.14	7.7(7)E	0	7.5(1.4)	0	7.5(1.4)	0.6(3)
^{105}Pd	22.33	5.5(3)	−2.6(1.6)	3.8(4)	0.8(1.0)	4.6(1.1)	20.(3.)
^{106}Pd	27.33	6.4(4)	0	5.1(6)	0	5.1(6)	0.304(29)
^{108}Pd	26.46	4.1(3)	0	2.1(3)	0	2.1(3)	8.5(5)
^{110}Pd	11.72	7.7(7)E	0	7.5(1.4)	0	7.5(1.4)	0.226(31)
Ag		5.922(7)		4.407(10)	0.58(3)	4.99(3)	63.3(4)
^{107}Ag	51.83	7.555(11)	1.00(13)	7.17(2)	0.13(3)	7.30(4)	37.6(1.2)
^{109}Ag	48.17	4.165(11)	−1.60(13)	2.18(1)	0.32(5)	2.50(5)	91.0(1.0)
Cd		4.87(5) − 0.70(1)i		3.04(6)	3.46(13)	6.50(12)	2520.(50.)
^{106}Cd	1.25	5.(2.)E	0	3.1(2.5)	0	3.1(2.5)	1

Table A1.1. (cont.)

Isotope	Natural abundance (%)	b_{coh} (fm)	b_{inc} (fm)	σ_{coh} (barn)	σ_{inc} (barn)	σ_{scat} (barn)	σ_{abs} (barn)
^{108}Cd	0.89	5.4(1)	0	3.7(1)	0	3.7(1)	1.1(3)
^{110}Cd	12.51	5.9(1)	0	4.4(1)	0	4.4(1)	11.(1.)
^{111}Cd	12.81	6.5(1)		5.3(2)	0.3(3)E	5.6(4)	24.(3.)
^{112}Cd	24.13	6.4(1)	0	5.1(2)	0	5.1(2)	2.2(5)
^{113}Cd	12.22	$-8.0(2)-5.73(11)i$		12.1(4)	0.3(3)E	12.4(5)	20600.(400.)
^{114}Cd	28.72	7.5(1)	0	7.1(2)	0	7.1(2)	0.34(2)
^{116}Cd	7.47	6.3(1)	0	5.0(2)	0	5.0(2)	0.075(13)
In		$4.065(20)-0.0539(4)i$		2.08(2)	0.54(11)	2.62(11)	193.8(1.5)
^{113}In	4.3	5.39(6)	±0.017(1)	3.65(8)	0.000037(5)	3.65(8)	12.0(1.1)
^{115}In	95.7	$4.01(2)-0.0562(6)i$	-2.1(2)	2.02(2)	0.55(11)	2.57(11)	202.(2.)
Sn		6.225(2)		4.870(3)	0.022(5)	4.892(6)	0.626(9)
^{112}Sn	1.0	6.(1.)E	0	4.5(1.5)	0	4.5(1.5)	1.00(11)
^{114}Sn	0.7	6.2(3)	0	4.8(5)	0	4.8(5)	0.114(30)
^{115}Sn	0.4	6.(1.)E		4.5(1.5)	0.3(3)E	4.8(1.5)	30.(7.)
^{116}Sn	14.7	5.93(5)	0	4.42(7)	0	4.42(7)	0.14(3)
^{117}Sn	7.7	6.48(5)		5.28(8)	0.3(3)E	5.6(3)	2.3(5)
^{118}Sn	24.3	6.07(5)	0	4.63(8)	0	4.63(8)	0.22(5)
^{119}Sn	8.6	6.12(5)		4.71(8)	0.3(3)E	5.0(3)	2.2(5)
^{120}Sn	32.4	6.49(5)	0	5.29(8)	0	5.29(8)	0.14(3)
^{122}Sn	4.6	5.74(5)	0	4.14(7)	0	4.14(7)	0.18(2)
^{124}Sn	5.6	5.97(5)	0	4.48(8)	0	4.48(8)	0.133(5)
Sb		5.57(3)		3.90(4)	0.00(7)	3.90(6)	4.91(5)
^{121}Sb	57.3	5.71(6)	-0.05(15)	4.10(9)	0.0003(19)	4.10(9)	5.75(12)
^{123}Sb	42.7	5.38(7)	-0.10(15)	3.64(9)	0.001(4)	3.64(9)	3.8(2)

Te		5.80(3)		4.23(4)	0.09(6)	4.32(5)	4.7(1)
120Te	0.096	5.3(5)	0	3.5(7)	0	3.5(7)	2.3(3)
122Te	2.60	3.8(2)	0	1.8(2)	0	1.8(2)	3.4(5)
123Te	0.908	−0.05(25) − 0.116(8)i	−2.04(9)	0.002(3)	0.52(5)	0.52(5)	418.(30.)
124Te	4.816	7.96(10)	0	8.0(2)	0	8.0(2)	6.8(1.3)
125Te	7.14	5.02(8)	−0.26(13)	3.17(10)	0.008(8)	3.18(10)	1.55(16)
126Te	18.95	5.56(7)	0	3.88(10)	0	3.88(10)	1.04(15)
128Te	31.69	5.89(7)	0	4.36(10)	0	4.36(10)	0.215(8)
130Te	33.80	6.02(7)	0	4.55(11)	0	4.55(11)	0.29(6)
127I	100	5.28(2)	1.58(15)	3.50(3)	0.31(6)	3.81(7)	6.15(6)
Xe		4.92(3)		3.04(4)			23.9(1.2)
124Xe	0.10		0		0		165.(20.)
126Xe	0.09		0		0		3.5(8)
128Xe	1.91		0		0		<8
129Xe	26.4						21.(5.)
130Xe	4.1		0		0		<26
131Xe	21.2						85.(10.)
132Xe	26.9		0		0		0.45(6)
134Xe	10.4		0		0		0.265(20)
136Xe	8.9		0		0		0.26(2)
133Cs	100	5.42(2)	1.29(15)	3.69(3)	0.21(5)	3.90(6)	29.0(1.5)
Ba		5.07(3)		3.23(4)	0.15(11)	3.38(10)	1.1(1)
130Ba	0.11	−3.6(6)	0	1.6(5)	0	1.6(5)	30.(5.)
132Ba	0.10	7.8(3)	0	7.6(6)	0	7.6(6)	7.0(8)
134Ba	2.42	5.7(1)	0	4.08(14)	0	4.08(14)	2.0(1.6)

Table A1.1. (*cont.*)

Isotope	Natural abundance (%)	b_{coh} (fm)	b_{inc} (fm)	σ_{coh} (barn)	σ_{inc} (barn)	σ_{scat} (barn)	σ_{abs} (barn)
135Ba	6.59	4.67(10)		2.74(12)	0.5(5)E	3.2(5)	5.8(9)
136Ba	7.85	4.91(8)	0	3.03(10)	0	3.03(10)	0.68(17)
137Ba	11.23	6.83(10)		5.86(17)	0.5(5)E	6.4(5)	3.6(2)
138Ba	71.70	4.84(8)	0	2.94(10)	0	2.94(10)	0.27(14)
La		8.24(4)		8.53(8)	1.13(19)	9.66(17)	8.97(4)
138La	0.09	8.(2.)E		8.(4.)	0.5(5)E	8.5(4.0)	57.(6.)
139La	99.91	8.24(4)	3.0(2)	8.53(8)	1.13(15)	9.66(17)	8.93(4)
Ce		4.84(2)		2.94(2)	0.00(10)	2.94(10)	0.63(4)
136Ce	0.19	5.80(9)	0	4.23(13)	0	4.23(13)	7.3(1.5)
138Ce	0.25	6.70(9)	0	5.64(15)	0	5.64(15)	1.1(3)
140Ce	88.48	4.84(9)	0	2.94(11)	0	2.94(11)	0.57(4)
142Ce	11.08	4.75(9)	0	2.84(11)	0	2.84(11)	0.95(5)
141Pr	100	4.58(5)	−0.35(3)	2.64(6)	0.015(3)	2.66(6)	11.5(3)
Nd		7.69(5)		7.43(10)	9.2(8)	16.6(8)	50.5(1.2)
142Nd	27.16	7.7(3)	0	7.5(6)	0	7.5(6)	18.7(7)
143Nd	12.18	14.(2.)E	±21.(1.)	25.(7.)	55.(7.)	80.(2.)	334.(10.)
144Nd	23.80	2.8(3)	0	1.0(2)	0	1.0(2)	3.6(3)
145Nd	8.29	14.(2.)E		25.(7.)	5.(5.)E	30.(9.)	42.(2.)
146Nd	17.19	8.7(2)	0	9.5(4)	0	9.5(4)	1.4(1)
148Nd	5.75	5.7(3)	0	4.1(4)	0	4.1(4)	2.5(2)
150Nd	5.63	5.3(2)	0	3.5(3)	0	3.5(3)	1.2(2)
147Pm	(2.62 a)	12.6(4)	±3.2(2.5)	20.0(1.3)	1.3(2.0)	21.3(1.5)	168.4(3.5)
Sm		0.80(2) − 1.65(2)i −3.(4.)E		0.422(9)	39.(3.)	39.(3.)	5922.(56.)
144Sm	3.1		0	1.(3.)	0	1.(3.)	0.7(3)
147Sm	15.1	14.(3.)	±11.(7.)	25.(11.)	14.(19.)	39.(16.)	57.(3.)

148 Sm	11.3	−3.(4.)E	0	0	1.(3.)	1.(3.)	2.4(6)
149 Sm	13.9	−19.2(1) − 11.7(1)i	±31.4(6) − 10.3(1)i	137.(5.)	63.5(6)	200.(5.)	42080.(400.)
150 Sm	7.4	14.(3.)	0	0	25.(11.)	25.(11.)	104.(4.)
152 Sm	26.6	−5.0(6)	0	0	3.1(8)	3.1(8)	206.(6.)
154 Sm	22.6	9.3(1.0)	0	0	11.(2.)	11.(2.)	8.4(5)
Eu		7.22(2) − 1.26(1)i		2.5(4)	6.75(4)	9.2(4)	4530.(40.)
151 Eu	47.8	6.13(14) − 2.53(3)i	±4.5(4) − 2.14(2)i	3.1(4)	5.5(2)	8.6(4)	9100.(100.)
153 Eu	52.2	8.22(12)	±3.2(9)	1.3(7)	8.5(2)	9.8(7)	312.(7.)
Gd		6.5(5) − 13.82(3)i		151.(2.)	29.3(8)	180.(2.)	49700.(125.)
152 Gd	0.2	10.(3.)E	0	0	13.(8.)	13.(8.)	735.(20.)
154 Gd	2.1	10.(3.)E	0	0	13.(8.)	13.(8.)	85.(12.)
155 Gd	14.8	6.0(1) − 17.0(1)i	±5.(5.)E − 13.16(9)i	25.(6.)	40.8(4)	66.(6.)	61100.(400.)
156 Gd	20.6	6.3(4)	0	0	5.0(6)	5.0(6)	1.5(1.2)
157 Gd	15.7	−1.14(2) − 71.9(2)i	±5.(5.)E − 55.8(2)i	394.(7.)	650.(4.)	1044.(8.)	259000.(700.)
158 Gd	24.8	9.(2.)	0	0	10.(5.)	10.(5.)	2.2(2)
160 Gd	21.8	9.15(5)	0	0	10.52(11)	10.52(11)	0.77(2)
159 Tb	100	7.38(3)	−0.17(7)	0.004(3)	6.84(6)	6.84(6)	23.4(4)
Dy		16.9(2) − 0.276(4)i		54.4(1.2)	35.9(8)	90.3(9)	994.(13.)
156 Dy	0.06	6.1(5)	0	0	4.7(8)	4.7(8)	33.(3.)
158 Dy	0.10	6.(4.)E	0	0	5.(6.)	5.(6.)	43.(6.)
160 Dy	2.34	6.7(4)	0	0	5.6(7)	5.6(7)	56.(5.)
161 Dy	19.0	10.3(4)	±4.9(8)	3.(1.)	13.3(1.0)	16.(1.)	600.(25.)
162 Dy	25.5	−1.4(5)	0	0	0.25(18)	0.25(18)	194.(10.)
163 Dy	24.9	5.0(4)	1.3(3)	0.21(10)	3.1(5)	3.3(5)	124.(7.)
164 Dy	28.1	49.4(2) − 0.79(1)i	0	0	307.(3.)	307.(3.)	2840.(40.)

Table A1.1. (cont.)

Isotope	Natural abundance (%)	b_{coh} (fm)	b_{inc} (fm)	σ_{coh} (barn)	σ_{inc} (barn)	σ_{scat} (barn)	σ_{abs} (barn)
^{165}Ho	100	8.01(8)	−1.70(8)	8.06(16)	0.36(3)	8.42(16)	64.7(1.2)
Er		7.79(2)		7.63(4)	1.1(3)	8.7(3)	159.(4.)
^{162}Er	0.14	8.8(2)	0	9.7(4)	0	9.7(4)	19.(2.)
^{164}Er	1.56	8.2(2)	0	8.4(4)	0	8.4(4)	13.(2.)
^{166}Er	33.4	10.6(2)	0	14.1(5)	0	14.1(5)	19.6(1.5)
^{167}Er	22.9	3.0(3)	1.0(3)	1.1(2)	0.13(8)	1.2(2)	659.(16.)
^{168}Er	27.1	7.4(4)	0	6.9(7)	0	6.9(7)	2.74(8)
^{170}Er	14.9	9.6(5)	0	11.6(1.2)	0	11.6(1.2)	5.8(3)
^{169}Tm	100	7.07(3)	0.9(3)	6.28(5)	0.10(7)	6.38(9)	100.(2.)
Yb		12.43(3)		19.42(9)	4.0(2)	23.4(2)	34.8(8)
^{168}Yb	0.14	−4.07(2) − 0.62(1)i	0	2.13(2)	0	2.13(2)	2230.(40.)
^{170}Yb	3.06	6.77(10)	0	5.8(2)	0	5.8(2)	11.4(1.0)
^{171}Yb	14.3	9.66(10)	−5.59(17)	11.7(2)	3.9(2)	15.6(3)	48.6(2.5)
^{172}Yb	21.9	9.43(10)	0	11.2(2)	0	11.2(2)	0.8(4)
^{173}Yb	16.1	9.56(7)	−5.3(2)	11.5(2)	3.5(3)	15.0(4)	17.1(1.3)
^{174}Yb	31.8	19.3(1)	0	46.8(5)	0	46.8(5)	69.4(5.0)
^{176}Yb	12.7	8.72(10)	0	9.6(2)	0	9.6(2)	2.85(5)
Lu		7.21(3)		6.53(5)	0.7(4)	7.2(4)	74.(2.)
^{175}Lu	97.39	7.24(3)	±2.2(7)	6.59(5)	0.6(4)	7.2(4)	21.(3.)
^{176}Lu	2.61	6.1(1) − 0.57(1)i	±3.0(4) + 0.61(1)i	4.7(2)	1.2(3)	5.9(4)	2065.(35.)
Hf		7.77(14)		7.6(3)	2.6(5)	10.2(4)	104.1(5)
^{174}Hf	0.2	10.9(1.1)	0	15.(3.)	0	15.(3.)	561.(35.)
^{176}Hf	5.2	6.61(18)	0	5.5(3)	0	5.5(3)	23.5(3.1)
^{177}Hf	18.6	0.8(1.0)E	±0.9(1.3)	0.1(2)	0.1(3)	0.2(2)	373.(10.)
^{178}Hf	27.1	5.9(2)	0	4.4(3)	0	4.4(3)	84.(4.)

^{179}Hf	13.7	7.46(16)	±1.06(8)	7.0(3)	0.14(2)	7.1(3)	41.(3.)
^{180}Hf	35.2	13.2(3)	0	21.9(1.0)	0	21.9(1.0)	13.04(7)
Ta		6.91(7)		6.00(12)	0.01(17)	6.01(12)	20.6(5)
^{180}Ta	0.012	7.(2.)E	−0.29(3)	6.2(3.5)	0.5(5)E	7.(4.)	563.(60.)
^{181}Ta	99.988	6.91(7)		6.00(12)	0.011(2)	6.01(12)	20.5(5)
W		4.86(2)		2.97(2)	1.63(6)	4.60(6)	18.3(2)
^{180}W	0.1	5.(3.)E	0	3.(4.)	0	3.(4.)	30.(20.)
^{182}W	26.3	6.97(4)	0	6.10(7)	0	6.10(7)	20.7(5)
^{183}W	14.3	6.53(4)		5.36(7)	0.3(3)E	5.7(3)	10.1(3)
^{184}W	30.7	7.48(6)	0	7.03(11)	0	7.03(11)	1.7(1)
^{186}W	28.6	−0.72(4)	0	0.065(7)	0	0.065(7)	37.9(6)
Re		9.2(2)		10.6(5)	0.9(6)	11.5(3)	89.7(1.)
^{185}Re	37.40	9.0(3)	±2.0(1.8)	10.2(7)	0.5(9)	10.7(6)	112.(2.)
^{187}Re	62.60	9.3(3)	±2.8(1.1)	10.9(7)	1.0(8)	11.9(4)	76.4(1.0)
Os		10.7(2)		14.4(5)	0.3(8)	14.7(6)	16.0(4)
^{184}Os	0.02	10.(2.)E	0	13.(5.)	0	13.(5.)	3000.(150.)
^{186}Os	1.58	11.6(1.7)	0	17.(5.)	0	17.(5.)	80.(13.)
^{187}Os	1.6	10.(2.)E		13.(5.)	0.3(3)E	13.(5.)	320.(10.)
^{188}Os	13.3	7.6(3)	0	7.3(6)	0	7.3(6)	4.7(5)
^{189}Os	16.1	10.7(3)		14.4(8)	0.5(5)E	14.9(9)	25.(4.)
^{190}Os	26.4	11.0(3)	0	15.2(8)	0	15.2(8)	13.1(3)
^{192}Os	41.0	11.5(4)	0	16.6(1.2)	0	16.6(1.2)	2.0(1)
Ir		10.6(3)		14.1(8)	0.(3.)	14.(3.)	425.(2.)
^{191}Ir	37.3						954.(10.)
^{193}Ir	62.7						111.(5.)

Table A1.1. (*cont.*)

Isotope	Natural abundance (%)	b_{coh} (fm)	b_{inc} (fm)	σ_{coh} (barn)	σ_{inc} (barn)	σ_{scat} (barn)	σ_{abs} (barn)
Pt		9.60(1)		11.58(2)	0.13(11)	11.71(11)	10.3(3)
190Pt	0.01	9.0(1.0)	0	10.(2.)	0	10.(2.)	152.(4.)
192Pt	0.79	9.9(5)	0	12.3(1.2)	0	12.3(1.2)	10.0(2.5)
194Pt	32.9	10.55(8)	0	14.0(2)	0	14.0(2)	1.44(19)
195Pt	33.8	8.83(11)	−1.00(17)	9.8(2)	0.13(4)	9.9(2)	27.5(1.2)
196Pt	25.3	9.89(8)	0	12.3(2)	0	12.3(2)	0.72(4)
198Pt	7.2	7.8(1)	0	7.6(2)	0	7.6(2)	3.66(19)
197Au	100	7.63(6)	−1.84(10)	7.32(12)	0.43(5)	7.75(13)	98.65(9)
Hg		12.692(15)		20.24(5)	6.6(1)	26.8(1)	372.3(4.0)
196Hg	0.2	30.3(1.0)	0	115.(8.)	0	115.(8.)	3080.(180.)
198Hg	10.1		0		0		2.0(3)
199Hg	17.0	16.9(4)	±15.5	36.(2.)	30.(3.)	66.(2.)	2150.(48.)
200Hg	23.1		0		0		< 60
201Hg	13.2						7.8(2.0)
202Hg	29.6		0		0		4.89(5)
204Hg	6.8		0		0		0.43(10)
Tl		8.776(5)		9.678(11)	0.21(15)	9.89(15)	3.43(6)
203Tl	29.524	6.99(16)	1.06(14)	6.14(28)	0.14(4)	6.28(28)	11.4(2)
205Tl	70.476	9.52(7)	−0.242(17)	11.39(17)	0.007(1)	11.40(17)	0.104(17)
Pb		9.405(3)		11.115(7)	0.0030(7)	11.118(7)	0.171(2)
204Pb	1.4	9.90(10)	0	12.3(2)	0	12.3(2)	0.65(7)
206Pb	24.1	9.22(5)	0	10.68(12)	0	10.68(12)	0.0300(8)
207Pb	22.1	9.28(4)	0.14(6)	10.82(9)	0.002(2)	10.82(9)	0.699(10)
208Pb	52.4	9.50(2)	0	11.34(5)	0	11.34(5)	0.00048(3)

Isotope							
^{209}Bi	100	8.532(2)	0.259(15)	9.148(4)	0.0084(10)	9.156(4)	0.0338(7)
Po							
At							
Rn							
Fr							
^{226}Ra	$(1.60 \times 10^3$ a$)$	10.0(1.0)	0	13.(3.)	0	13.(3.)	12.8(1.5)
Ac							·	
^{232}Th	100	10.31(3)	0	13.36(8)	0	13.36(8)	7.37(6)
Pa	$(3.28 \times 10^4$ a$)$	9.1(3)		10.4(7)	0.1(3.3)	10.5(3.2)	200.6(2.3)
U		8.417(5)		8.903(11)	0.005(16)	8.908(11)	7.57(2)
^{233}U	$(1.59 \times 10^5$ a$)$	10.1(2)	±1.(3.)	12.8(5)	0.1(6)	12.9(3)	574.7(1.0)
^{234}U	0.005	12.4(3)	0	19.3(9)	0	19.3(9)	100.1(1.3)
^{235}U	0.720	10.47(4)	±1.3(6)	13.78(11)	0.2(2)	14.0(2)	680.9(1.1)
^{238}U	99.275	8.402(5)	0	8.871(11)	0	8.871(11)	2.68(2)
^{237}Np	$(2.14 \times 10^6$ a$)$	10.55(10)		14.0(3)	0.5(5)E	14.5(6)	175.9(2.9)
^{238}Pu	$(87.74$ a$)$	14.1(5)	0	25.0(1.8)	0	25.0(1.8)	558.(7.)
^{239}Pu	$(2.41 \times 10^4$ a$)$	7.7(1)	±1.3(1.9)	7.5(2)	0.2(6)	7.7(6)	1017.3(2.1)
^{240}Pu	$(6.56 \times 10^3$ a$)$	3.5(1)	0	1.54(9)	0	1.54(9)	289.6(1.4)
^{242}Pu	$(3.76 \times 10^5$ a$)$	8.1(1)	0	8.2(2)	0	8.2(2)	18.5(5)
^{243}Am	$(7.37 \times 10^3$ a$)$	8.3(2)	±2.(7.)	8.7(4)	0.3(2.6)	9.0(2.6)	75.3(1.8)
^{244}Cm	$(18.10$ a$)$	9.5(3)	0	11.3(7)	0	11.3(7)	16.2(1.2)
^{246}Cm	$(4.7 \times 10^3$ a$)$	9.3(2)	0	10.9(5)	0	10.9(5)	1.36(17)
^{248}Cm	$(3.5 \times 10^5$ a$)$	7.7(2)	0	7.5(4)	0	7.5(4)	3.00(26)

Source: Sears, V. F. (1992) *Neutron News* **3**(3), 303.

Appendix 2

Crystallographic data

Table A2.1. *Crystallographic data for selected elements at room temperature (except as noted).*

Formula	Crystal system	Space group	Lattice constants a (Å)	c (Å)
Al	Cubic	$Fm3m$	4.04964	
Be	Hexagonal	$P6_3/mmc$	2.2854	3.5807
C (diamond)	Cubic	$Fd3m$	3.5667	
C (graphite)	Hexagonal	$P6_3/mmc$	2.4612(1)	6.7079(7)
Co	Hexagonal	$P6_3/mmc$	2.5074	4.0699
Cr	Cubic	$Im3m$	2.8845(5)	
Cu	Cubic	$Fm3m$	3.61509(4)	
Dy	Hexagonal	$P6_3/mmc$	3.5903(1)	5.6475(2)
Er	Hexagonal	$P6_3/mmc$	3.5588(3)	5.5874(3)
Fe	Cubic	$Im3m$	2.86645(1)	
Ge	Cubic	$Fd3m$	5.657764(10)	
Ho	Hexagonal	$P6_3/mmc$	3.5773(1)	5.6158(2)
K	Cubic	$Im3m$	5.344(5)	
^{7}Li	Cubic	$Im3m$	3.5092(6)	
Mg	Hexagonal	$P6_3/mmc$	3.20939(3)	5.21053(5)
Mo	Cubic	$Im3m$	3.1472	
N_2 ($T = 50$ K)	Hexagonal	$P6_3/mmc$	3.93(16)	6.50(51)
Nb (H_2 free)	Cubic	$Im3m$	3.3008(3)	
Ni	Cubic	$Fm3m$	3.52394(8)	
Pb	Cubic	$Fm3m$	4.9505	
Pd	Cubic	$Fm3m$	3.8898	
Rb	Cubic	$Im3m$	5.709	
Rh	Cubic	$Fm3m$	3.8043(3)	
Si	Cubic	$Fd3m$	5.43072(5)	
Ta	Cubic	$Im3m$	3.3058	
Tb	Hexagonal	$P6_3/mmc$	3.6010(3)	5.6936(2)
V	Cubic	$Im3m$	3.0399(3)	
W	Cubic	$Im3m$	3.16517	
Zn	Hexagonal	$P6_3/mmc$	2.6589	4.9349

Source: Donnay, J. D. H., Mason, W. P., and Wood, E. A. (1972). In *American Institute of Physics Handbook*, ed. D. E. Gray (McGraw-Hill, New York), p. **9–2**.

Table A2.2. *Crystallographic data for common compounds at room temperature.*

Formula	Crystal system	Space group	Lattice constants a, c (Å)
AgCl	Cubic	$Fm3m$	5.5491
AgBr	Cubic	$Fm3m$	5.7745
Al_2O_3 (α)	Rhombohedral	$R\bar{3}c$	4.759216, 12.99127
$BaTiO_3$	Tetragonal	$P4/mmm$	3.9939, 4.0346
$CoFe_2O_4$	Cubic	$Fd3m$	8.429
Cr_2O_3	Hexagonal	$R\bar{3}c$	5.35, $\alpha = 54°58'$
Fe_2O_3	Hexagonal	$R\bar{3}c$	5.42, $\alpha = 55°17'$
Fe_3O_4	Cubic	$Fd3m$	8.396
KBr	Cubic	$Fm3m$	6.5982
KCl	Cubic	$Fm3m$	6.29294
KF	Cubic	$Fm3m$	5.347
KI	Cubic	$Fm3m$	7.06555
LiBr	Cubic	$Fm3m$	5.5013
LiCl	Cubic	$Fm3m$	5.13988
LiF	Cubic	$Fm3m$	4.0262
KH_2PO_4	Tetragonal	$I\bar{4}2d$	7.448, 6.977
$LaAlO_3$	Cubic	$Pm3m$	3.78
$MgAl_2O_4$	Cubic	$Fd3m$	8.0800
MgO	Cubic	$Fm3m$	4.213
MnF_2	Tetragonal	$P4_2/mnm$	
MnO	Cubic	$Fm3m$	4.435
MnSb	Hexagonal	$P6_3/mmc$	4.120, 5.784
MnSi	Cubic	$P2_13$	4.548
NaCl	Cubic	$Fm3m$	5.6402
$NaClO_3$	Cubic	$P2_13$	6.568
NaF	Cubic	$Fm3m$	4.6342
NiAs	Hexagonal	$P6_3/mmc$	3.638, 5.059
$NiFe_2O_4$	Cubic	$Fd3m$	8.339
$NH_4H_2PO_4$	Tetragonal	$I\bar{4}2d$	7.499, 7.548
RbF	Cubic	$Fm3m$	5.64 $\pm$ 0.02
SiC	Hexagonal	$P6_3mc$	3.076, 5.048
SiO_2 (α phase)	Hexagonal	$P3_121$ or $P3_221$	4.91343, 5.40506
SiO_2 (β phase)	Hexagonal	$P6_222$ or $P6_422$	5.01, 5.47
ZnO (18)	Hexagonal	$P6_3mc$	3.2427, 5.1948

Source: Donnay, J. D. H., Mason, W. P., and Wood, E. A. (1972). In *American Institute of Physics Handbook*, ed. D. E. Gray (McGraw-Hill, New York), p. **9–2**.

Appendix 3

Other useful tables

Table A3.1. *d-spacings and reciprocal spacings $d^* = 2\pi/d$ for common monochromator and analyzer crystals.*

Crystal reflection	d (Å)	$2\pi/d$ (Å^{-1})
PG(002)	3.35416	1.87325
PG(004)	1.67708	3.74650
Cu(111)	2.08717	3.01038
Cu(220)	1.27813	4.91593
Ge(111)	3.26627	1.92366
Ge(220)	2.00018	3.14131
Ge(311)	1.70576	3.68351
Ge(331)	1.29789	4.84107
Be(002)	1.79160	3.50702
Be(110)	1.14280	5.49806
Cu$_2$MnAl(111)	3.435	1.829

PG = pyrolytic graphite.

Table A3.2. Scattering angles ($2\theta_B$ in degrees) as a function of energy for a number of common monochromator crystals.

Energy (meV)	k (Å^{-1})	λ (Å)	PG (002)	Ge (111)	Cu (111)	Be (002)	Ge (311)	PG (004)	Cu (220)	Be (110)	Cu$_2$MnAl (111)
2.0	0.9824	6.3955	144.87	156.49							137.13
3.0	1.2032	5.2219	102.23	106.14							98.93
4.0	1.3894	4.5223	84.77	87.62							82.33
5.0	1.5534	4.0449	74.16	76.51	151.38						72.13
6.0	1.7016	3.6924	66.79	68.84	124.39						65.02
7.0	1.8380	3.4185	61.27	63.11	109.96	145.12					59.68
8.0	1.9649	3.1977	56.94	58.62	100.00	126.36	139.22	144.87			55.48
9.0	2.0841	3.0149	53.41	54.97	92.48	114.57	124.19	128.01			52.06
10.0	2.1968	2.8601	50.47	51.93	86.50	105.92	113.94	117.02			49.20
11.0	2.3040	2.7270	47.97	49.35	81.58	99.12	106.14	108.79			46.77
12.0	2.4065	2.6109	45.81	47.12	77.43	93.55	99.87	102.23			44.67
13.0	2.5047	2.5085	43.92	45.16	73.87	88.87	94.67	96.81	157.82		42.83
13.7	2.5713	2.4436	42.72	43.93	71.66	85.99	91.50	93.53	145.85		41.67
14.7	2.6635	2.3590	41.18	42.34	68.82	82.35	87.50	89.39	134.69		40.16
16.0	2.7788	2.2611	39.40	40.50	65.60	78.25	83.03	84.77	124.39	163.22	38.43
18.0	2.9473	2.1318	37.06	38.09	61.42	73.02	77.35	78.93	113.02	137.73	36.15
20.0	3.1068	2.0224	35.09	36.07	57.96	68.72	72.72	74.16	104.59	124.47	34.24
22.0	3.2584	1.9283	33.41	34.34	55.03	65.12	68.84	70.19	97.94	115.06	32.60
26.0	3.5422	1.7738	30.66	31.51	50.29	59.34	62.66	63.85	87.88	101.80	29.92
30.5	3.8366	1.6377	28.26	29.04	46.20	54.39	57.38	58.45	79.68	91.54	27.58
36.0	4.1681	1.5074	25.97	26.68	42.34	49.76	52.45	53.41	72.27	82.53	25.35
41.0	4.4482	1.4125	24.31	24.98	39.56	46.43	48.92	49.81	67.09	76.34	23.73
46.0	4.7116	1.3335	22.93	23.56	37.26	43.70	46.02	46.85	62.89	71.39	22.38
50.0	4.9122	1.2791	21.98	22.58	35.69	41.83	44.04	44.83	60.05	68.06	21.46

252

Table A3.2. (cont.)

Energy (meV)	k (Å^{-1})	λ (Å)	PG (002)	Ge (111)	Cu (111)	Be (002)	Ge (311)	PG (004)	Cu (220)	Be (110)	Cu$_2$MnAl (111)
55.0	5.1520	1.2196	20.95	21.52	33.97	39.80	41.89	42.64	56.99	64.50	20.45
60.0	5.3811	1.1676	20.05	20.59	32.49	38.04	40.03	40.74	54.36	61.44	19.57
65.0	5.6008	1.1218	19.25	19.78	31.18	36.49	38.40	39.08	52.06	58.79	18.79
70.0	5.8122	1.0810	18.55	19.05	30.02	35.12	36.95	37.60	50.04	56.46	18.11
75.0	6.0162	1.0444	17.91	18.40	28.98	33.89	35.65	36.28	48.23	54.38	17.49
80.0	6.2135	1.0112	17.34	17.81	28.04	32.78	34.48	35.09	46.60	52.52	16.93
85.0	6.4047	0.9810	16.82	17.27	27.18	31.78	33.42	34.01	45.14	50.84	16.42
90.0	6.5904	0.9534	16.34	16.78	26.40	30.86	32.46	33.03	43.80	49.31	15.95
95.0	6.7710	0.9280	15.90	16.33	25.69	30.02	31.57	32.12	42.57	47.91	15.52
100.0	6.9469	0.9045	15.50	15.92	25.03	29.24	30.75	31.29	41.44	46.62	15.13
110.0	7.2860	0.8624	14.77	15.17	23.84	27.85	29.28	29.80	39.43	44.33	14.42
120.0	7.6100	0.8257	14.14	14.52	22.82	26.64	28.01	28.50	37.69	42.35	13.80
130.0	7.9207	0.7933	13.58	13.95	21.91	25.58	26.89	27.36	36.16	40.62	13.26
140.0	8.2197	0.7644	13.09	13.44	21.10	24.64	25.90	26.35	34.80	39.08	12.78
150.0	8.5082	0.7385	12.64	12.98	20.38	23.79	25.00	25.44	33.58	37.70	12.34
160.0	8.7872	0.7150	12.24	12.57	19.73	23.02	24.20	24.62	32.49	36.46	11.95
170.0	9.0577	0.6937	11.87	12.19	19.13	22.33	23.46	23.87	31.49	35.34	11.59
180.0	9.3203	0.6741	11.54	11.85	18.59	21.69	22.79	23.19	30.58	34.31	11.26
190.0	9.5757	0.6562	11.23	11.53	18.09	21.10	22.18	22.56	29.75	33.37	10.96
200.0	9.8244	0.6395	10.94	11.24	17.63	20.56	21.61	21.98	28.98	32.50	10.68

PG = pyrolytic graphite.

Table A3.3. *Energy resolution (FWHM in meV), $\hbar\Delta_\omega$, see Eq. (4.29), for a constant-**Q** scan through a dispersionless surface at $\hbar\omega = 0$, with all collimations and monochromator and analyzer mosaics widths equal to 20'. Values are listed as a function of energy for a range of monochromator choices, with the analyzer assumed to be identical to the monochromator. Note that, for the specified conditions, the energy resolution is independent of Q.*

Energy (meV)	PG (002)	Ge (111)	Cu (111)	Be (002)	Ge (311)	PG (004)	Cu (220)	Be (110)	Cu₂MnAl (111)
2	0.007	0.005							0.009
3	0.028	0.026							0.030
4	0.051	0.049							0.053
5	0.077	0.074	0.015						0.080
6	0.106	0.102	0.037						0.110
7	0.138	0.133	0.057	0.026					0.142
8	0.172	0.166	0.078	0.047	0.035	0.029			0.177
9	0.208	0.201	0.100	0.067	0.055	0.051			0.214
10	0.247	0.239	0.124	0.088	0.076	0.071			0.254
11	0.288	0.279	0.148	0.109	0.096	0.092			0.296
12	0.330	0.320	0.174	0.131	0.117	0.113			0.340
13	0.375	0.364	0.201	0.154	0.139	0.134	0.030		0.386
14	0.422	0.409	0.229	0.178	0.162	0.157	0.056		0.433
15	0.470	0.456	0.259	0.203	0.186	0.180	0.078		0.483
20	0.736	0.715	0.420	0.340	0.316	0.308	0.180	0.123	0.756
25	1.039	1.009	0.605	0.497	0.465	0.454	0.290	0.225	1.066
30	1.374	1.336	0.810	0.672	0.631	0.617	0.412	0.334	1.410
35	1.740	1.692	1.035	0.863	0.812	0.795	0.546	0.453	1.784
40	2.133	2.074	1.276	1.069	1.008	0.987	0.690	0.580	2.187
50	2.995	2.914	1.807	1.522	1.438	1.410	1.007	0.862	3.070
60	3.950	3.843	2.396	2.025	1.917	1.880	1.360	1.175	4.048
70	4.988	4.854	3.038	2.574	2.438	2.392	1.745	1.517	5.112
80	6.105	5.941	3.728	3.164	2.999	2.944	2.161	1.887	6.256
100	8.551	8.323	5.243	4.460	4.232	4.155	3.076	2.700	8.762

PG = pyrolytic graphite.

Table A3.4. *Values of Bose factor for energy loss, $n(\omega) + 1$, tabulated for a number of energies and temperatures.*

Energy (meV)	Temperature (K)												
	0.3	1.5	4.5	10	20	30	50	70	100	150	200	300	400
0.1	1.02	1.86	4.40	9.13	17.74	26.35	43.59	60.82	86.67	129.76	172.85	259.02	345.19
0.5	1.00	1.02	1.38	2.27	3.97	5.69	9.13	12.57	17.74	26.35	34.97	52.21	69.44
1	1.00	1.00	1.08	1.46	2.27	3.12	4.83	6.55	9.13	13.43	17.74	26.35	34.97
2	1.00	1.00	1.01	1.11	1.46	1.86	2.69	3.54	4.83	6.98	9.13	13.43	17.74
3	1.00	1.00	1.00	1.03	1.21	1.46	1.99	2.55	3.40	4.83	6.26	9.13	12.00
5	1.00	1.00	1.00	1.00	1.06	1.17	1.46	1.77	2.27	3.12	3.97	5.69	7.41
7	1.00	1.00	1.00	1.00	1.02	1.07	1.25	1.46	1.80	2.39	3.00	4.22	5.44
10	1.00	1.00	1.00	1.00	1.00	1.02	1.11	1.24	1.46	1.86	2.27	3.12	3.97
15	1.00	1.00	1.00	1.00	1.00	1.00	1.03	1.09	1.21	1.46	1.72	2.27	2.83
20	1.00	1.00	1.00	1.00	1.00	1.00	1.01	1.04	1.11	1.27	1.46	1.86	2.27
30	1.00	1.00	1.00	1.00	1.00	1.00	1.00	1.01	1.03	1.11	1.21	1.46	1.72
40	1.00	1.00	1.00	1.00	1.00	1.00	1.00	1.00	1.01	1.05	1.11	1.27	1.46

Table A3.5. *Values of Bose factor for energy gain, n(ω), tabulated for a number of energies and temperatures.*

Energy (meV)	Temperature (K)												
	0.3	1.5	4.5	10	20	30	50	70	100	150	200	300	400
0.1	0.02	0.86	3.40	8.13	16.74	25.35	42.59	59.82	85.67	128.76	171.85	258.02	344.19
0.5	0.00	0.02	0.38	1.27	2.97	4.69	8.13	11.57	16.74	25.35	33.97	51.21	68.44
1	0.00	0.00	0.08	0.46	1.27	2.12	3.83	5.55	8.13	12.43	16.74	25.35	33.97
2	0.00	0.00	0.01	0.11	0.46	0.86	1.69	2.54	3.83	5.98	8.13	12.43	16.74
3	0.00	0.00	0.00	0.03	0.21	0.46	0.99	1.55	2.40	3.83	5.26	8.13	11.00
5	0.00	0.00	0.00	0.00	0.06	0.17	0.46	0.77	1.27	2.12	2.97	4.69	6.41
7	0.00	0.00	0.00	0.00	0.02	0.07	0.25	0.46	0.80	1.39	2.00	3.22	4.44
10	0.00	0.00	0.00	0.00	0.00	0.02	0.11	0.24	0.46	0.86	1.27	2.12	2.97
15	0.00	0.00	0.00	0.00	0.00	0.00	0.03	0.09	0.21	0.46	0.72	1.27	1.83
20	0.00	0.00	0.00	0.00	0.00	0.00	0.01	0.04	0.11	0.27	0.46	0.86	1.27
30	0.00	0.00	0.00	0.00	0.00	0.00	0.00	0.01	0.03	0.11	0.21	0.46	0.72
40	0.00	0.00	0.00	0.00	0.00	0.00	0.00	0.00	0.01	0.05	0.11	0.27	0.46

Appendix 4

The resolution function for a triple-axis neutron spectrometer

As discussed in Chap. 4, Cooper and Nathans (1967) showed that the resolution function can be expressed in the form

$$R(\omega - \omega_0, \mathbf{Q} - \mathbf{Q}_0) = R_0 \exp\left(-\tfrac{1}{2}\Delta\mathcal{Q}\mathrm{M}\Delta\mathcal{Q}\right), \qquad (A4.1)$$

where

$$\Delta\mathcal{Q} = \left(\frac{m_n}{\hbar Q_0}\Delta\omega, \Delta\mathbf{Q}\right) = \left(\frac{m_n}{\hbar Q_0}(\omega - \omega_0), \mathbf{Q} - \mathbf{Q}_0\right). \qquad (A4.2)$$

Here we have written the 4-vector $\Delta\mathcal{Q}$ in a form such that all components have the same dimension of inverse length. In terms of the single-axis resolution functions, $P_i(\mathbf{k}_i - \bar{\mathbf{k}}_i)$ and $P_f(\mathbf{k}_f - \bar{\mathbf{k}}_f)$, the resolution function is defined by

$$R(\Delta\omega, \Delta\mathbf{Q}) = \frac{\hbar^2}{m_n} \int d\mathbf{k}_i d\mathbf{k}_f P_i(\Delta\mathbf{k}_i) P_f(\Delta\mathbf{k}_f)\, \delta(\mathbf{Q} - \mathbf{k}_f + \mathbf{k}_i)$$

$$\times \delta[\omega - (\hbar/2m_n)(k_i^2 - k_f^2)]. \qquad (A4.3)$$

In order to evaluate this integral, we first need analytic forms for P_i and P_f. These are derived in §A4.1. A basic understanding of the shape of the resolution function can be obtained from a qualitative analysis of Eq. (A4.3), and such an analysis is presented in §A4.2. Finally, the resolution matrix M is derived in §A4.3.

Before beginning, we need to discuss conversions between different width conventions. The resolution function has a gaussian form, and the finite elements of a diagonalized resolution matrix correspond to the inverse squares of gaussian widths. Gaussian distributions will also be used throughout the derivations below. In practice, experimental parameters are usually specified in terms of the full-width at half-maximum (FWHM). For a given width parameter σ the conversion between FWHM and gaussian units is:

$$\sigma(\text{FWHM}) = \sigma(\text{gaussian}) \times \sqrt{8 \ln 2}, \qquad (A4.4)$$

$$= \sigma(\text{gaussian}) \times 2.3548. \qquad (A4.5)$$

In applying formulas, the appropriate form of the width parameters depends on the purpose. If one intends to calculate the full resolution function, then it is necessary to convert input parameters from FWHM to gaussian widths before evaluating the resolution matrix. Conversely, when evaluating a formula for, say, the resolution width along the energy axis, where one desires a result corresponding to the FWHM, values for the width parameters in the formula should also be in FWHM units.

A4.1 Single-axis resolution functions P_i and P_f

The discussion below assumes a conventional triple-axis spectrometer configuration as illustrated in Fig. 4.1.

A4.1.1 Derivation of P_i and P_f

To derive the function $P_i(\mathbf{k}_i - \overline{\mathbf{k}}_i)$ we start by considering transmission through the collimator C_0. This collimator defines an average direction along $\overline{\mathbf{k}}'_i$. Let $\mathbf{k}'_i$ be a wave vector close to $\overline{\mathbf{k}}'_i$; then we can write

$$\Delta \mathbf{k}'_i = \mathbf{k}'_i - \overline{\mathbf{k}}'_i = (\Delta k'_{i\parallel}, \Delta k'_{i\perp}, \Delta k'_{iz}), \qquad (A4.6)$$

where $\Delta k'_{i\parallel}$ is along $\overline{\mathbf{k}}'_i$, $\Delta k'_{i\perp}$ is the component perpendicular to $\overline{\mathbf{k}}'_i$ and in the (horizontal) scattering plane, and $\Delta k'_{iz}$ is the vertical component. The transmission function through C_0 is assumed to have the form (Sailor *et al.*, 1956)

$$T_0(\mathbf{k}'_i - \overline{\mathbf{k}}'_i) = \exp\left\{ -\frac{1}{2}\left[\left(\frac{\Delta k'_{i\perp}}{\alpha_0 \overline{k}'_i}\right)^2 + \left(\frac{\Delta k'_{iz}}{\beta_0 \overline{k}'_i}\right)^2 \right] \right\}, \qquad (A4.7)$$

where the gaussian angular width of the transmission function is α_0 in the horizontal plane, and β_0 in the vertical direction. Similarly, the transmission function for collimator C_1 is

$$T_1(\mathbf{k}_i - \overline{\mathbf{k}}_i) = \exp\left\{ -\frac{1}{2}\left[\left(\frac{\Delta k_{i\perp}}{\alpha_1 \overline{k}_i}\right)^2 + \left(\frac{\Delta k_{iz}}{\beta_1 \overline{k}_i}\right)^2 \right] \right\}. \qquad (A4.8)$$

The monochromator crystal may have a mosaic distribution of gaussian width η_M in the horizontal direction and η'_M in the vertical. Let a given crystallite be rotated from the mean plane of the monochromator by angles γ_M and δ_M in the horizontal and vertical directions, respectively. Then the probability of a crystallite having orientation (γ_M, δ_M) is

$$P_M(\gamma_M, \delta_M) = \frac{1}{2\pi\eta_M\eta'_M} \exp\left\{-\frac{1}{2}\left[\left(\frac{\gamma_M}{\eta_M}\right)^2 + \left(\frac{\delta_M}{\eta'_M}\right)^2\right]\right\}. \tag{A4.9}$$

The reflectivity of the monochromator, $R_M(k_i)$, is a function of the wave vector. The precise functional form will depend on the nature of the crystal (Riste and Otnes, 1969; Dorner, 1971).

The single-axis resolution function is obtained by multiplying together the transmission functions for the collimators and the monochromator mosaic distribution, and then integrating over all incident wave vectors $\mathbf{k}'_i$:

$$P_i(\mathbf{k}_i - \overline{\mathbf{k}}_i) = R_M(\overline{k}_i) \int d\mathbf{k}'_i \, T_0(\mathbf{k}'_i - \overline{\mathbf{k}}'_i) P_M(\gamma_M, \delta_M) T_1(\mathbf{k}_i - \overline{\mathbf{k}}_i). \tag{A4.10}$$

Before we can evaluate the integrals in this equation, we must consider the constraints on the wave vectors due to the diffraction process. Elastic scattering requires $k'_i = k_i$, and taking $\overline{k}'_i = \overline{k}_i$, this implies $\Delta k'_{i\parallel} = \Delta k_{i\parallel}$. The relationships between transverse components are most easily evaluated by considering the Laue condition for diffraction (see Fig. A4.1)

$$\mathbf{k}_i - \mathbf{k}'_i = \mathbf{G}_M, \tag{A4.11}$$

where $\mathbf{G}_M$ is a reciprocal-lattice vector perpendicular to the lattice planes responsible for diffraction by the monochromator crystal, and

$$G_M = 2\overline{k}_i \sin\theta_M. \tag{A4.12}$$

Equivalently, we can write Eq. (A4.11) as

$$\Delta\mathbf{k}_i - \Delta\mathbf{k}'_i = \Delta\mathbf{G}_M, \tag{A4.13}$$

where $\Delta\mathbf{G}_M$ is the difference from the average $\mathbf{G}_M$ for a crystallite rotated by the angles γ_M and δ_M. In a coordinate system relative to the average crystal position, with y along $\mathbf{G}_M$, x in the horizontal plane and z vertical, one finds

$$\Delta\mathbf{k}_i = \begin{pmatrix} \Delta k_{i\parallel} \cos\theta_M - \Delta k_{i\perp} \sin\theta_M \\ \Delta k_{i\parallel} \sin\theta_M + \Delta k_{i\perp} \cos\theta_M \\ \Delta k_{iz} \end{pmatrix}, \tag{A4.14}$$

$$\Delta\mathbf{k}'_i = \begin{pmatrix} \Delta k'_{i\parallel} \cos\theta_M + \Delta k'_{i\perp} \sin\theta_M \\ -\Delta k'_{i\parallel} \sin\theta_M + \Delta k'_{i\perp} \cos\theta_M \\ \Delta k'_{iz} \end{pmatrix}, \tag{A4.15}$$

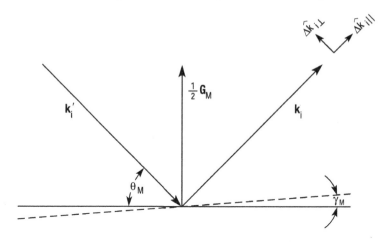

Fig. A4.1. Sketch of the diffraction condition for the monochromator.

and

$$\Delta \mathbf{G}_M = \begin{pmatrix} \gamma_M G_M \\ 0 \\ \delta_M G_M \end{pmatrix}. \tag{A4.16}$$

Plugging these results into Eq. (A4.13) we obtain

$$\gamma_M = -(\Delta k_{i\perp} + \Delta k'_{i\perp})/2\overline{k}_i, \tag{A4.17}$$

$$\delta_M = (\Delta k_{iz} - \Delta k'_{iz})/2\overline{k}_i \sin \theta_M, \tag{A4.18}$$

$$\Delta k'_{i\perp} = \Delta k_{i\perp} + 2\Delta k_{i\parallel} \tan \theta_M. \tag{A4.19}$$

Equations (A4.17) and (A4.18) can be used to substitute for γ_M and δ_m in Eq. (A4.10), while the conditions $\Delta k'_{i\parallel} = \Delta k_{i\parallel}$ and Eq. (A4.19) can be included in the integrals as δ functions.

After evaluating the integrals in Eq. (A4.10), we can write the result in the form

$$P_i(\mathbf{k}_i - \overline{\mathbf{k}}_i) = R_M(\overline{k}_i)P_{i0} \exp\left(-\tfrac{1}{2}\Delta\hat{\mathbf{k}}_i A_i \Delta\hat{\mathbf{k}}_i\right), \tag{A4.20}$$

where

$$\Delta\hat{\mathbf{k}}_i = (\mathbf{k}_i - \overline{\mathbf{k}}_i)/\overline{k}_i, \tag{A4.21}$$

and

$$P_{i0} = \frac{G_M}{\sqrt{2\pi}\eta_M} \left(\frac{\beta_0^2}{\beta_0^2 + 4\eta_M'^2 \sin^2 \theta_M}\right)^{\frac{1}{2}}. \tag{A4.22}$$

The 3×3 matrix A_i has the form

$$A_i = \begin{pmatrix} a_{i,11} & a_{i,12} & 0 \\ a_{i,12} & a_{i,22} & 0 \\ 0 & 0 & a_{i,33} \end{pmatrix} \qquad (A4.23)$$

with

$$a_{i,11} = \tan^2 \theta_M \left(\frac{4}{\alpha_0^2} + \frac{1}{\eta_M^2} \right), \qquad (A4.24)$$

$$a_{i,12} = \varepsilon_M |\tan \theta_M| \left(\frac{2}{\alpha_0^2} + \frac{1}{\eta_M^2} \right), \qquad (A4.25)$$

$$a_{i,22} = \frac{1}{\alpha_0^2} + \frac{1}{\alpha_1^2} + \frac{1}{\eta_M^2}, \qquad (A4.26)$$

$$a_{i,33} = \frac{1}{\beta_1^2} + \frac{1}{\beta_0^2 + 4\eta_M'^2 \sin^2 \theta_M}. \qquad (A4.27)$$

In Eq. (A4.25) above, we have introduced the parameter ε_M ($= \pm 1$) to explicitly account for the sign of θ_M. In the case of the conventional "W" configuration illustrated for a left-handed system in Fig. A4.2, the monochromator scattering angle $2\theta_M$ increases in a clockwise sense, and hence $\varepsilon_M = -1$. In contrast, the scattering angle at the sample position, $2\theta_S$, has a counterclockwise sense, so that $\varepsilon_S = +1$. The sign parameters ε_i determine the orientation of the resolution function in scattering space.

The resolution function $P_f(\mathbf{k}_f - \overline{\mathbf{k}}_f)$ for the analyzing section of the spectrometer is derived in a similar way. The wave vector of a neutron scattered by the sample is $\mathbf{k}_f$, and after diffraction by the analyzer it becomes $\mathbf{k}_f'$. The only difference from the previous derivation is that here one integrates over all wave vectors $\mathbf{k}_f'$ exiting the final collimator C_3 rather than over the wave vectors $\mathbf{k}_i'$ entering the initial collimator C_0. The relevant formulas are

$$P_f(\mathbf{k}_f - \overline{\mathbf{k}}_f) = R_A(\overline{k}_f) P_{f0} \exp \left(-\tfrac{1}{2} \Delta \hat{\mathbf{k}}_f A_f \Delta \hat{\mathbf{k}}_f \right), \qquad (A4.28)$$

with

$$P_{f0} = \frac{G_A}{\sqrt{2\pi}\eta_A} \left(\frac{\beta_3^2}{\beta_3^2 + 4\eta_A'^2 \sin^2 \theta_A} \right)^{\frac{1}{2}}, \qquad (A4.29)$$

and

$$a_{f,11} = \tan^2 \theta_A \left(\frac{4}{\alpha_3^2} + \frac{1}{\eta_A^2} \right), \qquad (A4.30)$$

$$a_{f,12} = -\varepsilon_A |\tan \theta_A| \left(\frac{2}{\alpha_3^2} + \frac{1}{\eta_A^2} \right), \qquad (A4.31)$$

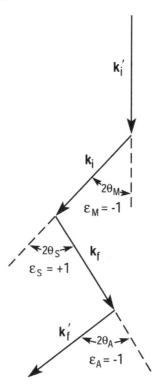

Fig. A4.2. Diagram illustrating the sign convention for the scattering angles.

$$a_{f,22} = \frac{1}{\alpha_3^2} + \frac{1}{\alpha_2^2} + \frac{1}{\eta_A^2}, \tag{A4.32}$$

$$a_{f,33} = \frac{1}{\beta_2^2} + \frac{1}{\beta_3^2 + 4\eta_A'^2 \sin^2 \theta_A}. \tag{A4.33}$$

Note the explicit sign change in the formula $a_{f,12}$ relative to $a_{i,12}$.

A4.1.2 Properties of P_i and P_f

From Eq. (A4.20), the points $\mathbf{k}_i$ for which $P_i = \frac{1}{2}$ are determined by the equation

$$\Delta \hat{\mathbf{k}}_i A_i \Delta \hat{\mathbf{k}}_i = 2 \ln 2. \tag{A4.34}$$

This defines a three-dimensional ellipsoid for which $\hat{\mathbf{k}}_{iz}$ is one of the principal axes. In the horizontal plane the principal axes are rotated with respect to the axes $\hat{\mathbf{k}}_{i\parallel}$ and $\hat{\mathbf{k}}_{i\perp}$ by an angle Θ_i as indicated in Fig. A4.3. For $\alpha_0 = \alpha_1$,

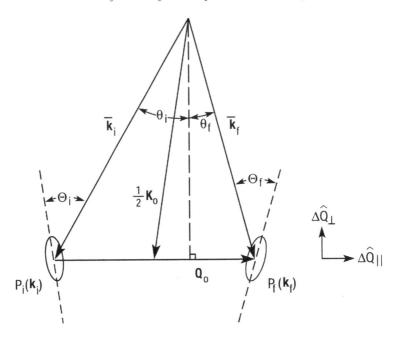

Fig. A4.3. Diagram illustrating the orientation of the ellipsoids for P_i and P_f with respect to $\mathbf{k}_i$, $\mathbf{k}_f$, and $\mathbf{Q}_0$ for the left-handed W configuration shown in the previous figure. $\mathbf{K}_0$ is defined by Eq. (A4.51).

Θ_i is given by

$$\tan 2\Theta_i = \frac{-2\varepsilon_M |\tan \theta_M|}{1 - (1 + B_i) \tan^2 \theta_M}, \tag{A4.35}$$

where

$$B_i = 2\eta_M^2 / (\alpha_0^2 + 2\eta_M^2). \tag{A4.36}$$

For $\eta_M = 0$, the ellipse in the horizontal plane collapses to a line with $\Theta_i = -\varepsilon_M |\theta_M|$. For finite η_M the ellipse rotates to $|\Theta_i| < |\theta_M|$, but for $\tan^2 \theta_M \ll 1$, the rotation is very small. The ellipsoid corresponding to P_f is rotated in the opposite direction, with $\Theta_f \approx +\varepsilon_A |\theta_A|$.

The total neutron flux through the monochromator arm of the spectrometer is proportional to the "volume" V_i of the single-axis resolution function given by

$$V_i = \int d\mathbf{k}_i P_i(\mathbf{k}_i - \overline{\mathbf{k}}_i). \tag{A4.37}$$

Using Eqs. (A4.20)–(A4.27) one finds

$$V_i = \frac{\bar{k}_i^3}{\tan\theta_{\mathrm{M}}} R_{\mathrm{M}}(\bar{k}_i) V_i', \qquad (A4.38)$$

where

$$V_i' = 2\pi G_{\mathrm{M}} \left(\frac{\alpha_0^2 \alpha_1^2}{\alpha_0^2 + \alpha_1^2 + 4\eta_{\mathrm{M}}^2} \right)^{\frac{1}{2}} \left(\frac{\beta_0^2 \beta_1^2}{\beta_0^2 + \beta_1^2 + 4\eta_{\mathrm{M}}'^2 \sin^2\theta_{\mathrm{M}}} \right)^{\frac{1}{2}}. \qquad (A4.39)$$

A symmetrical result is obtained for V_f. These formulas agree to within a constant factor with the result obtained by Dorner (1972). Note that the neutron flux through the monochromator arm is proportional to $k_i^3/\tan\theta_{\mathrm{M}}$, while the flux through the analyzer arm is proportional to $k_f^3/\tan\theta_{\mathrm{A}}$.

A4.2 Qualitative analysis of the resolution function

To obtain the triple-axis resolution function, we have to find all of the possible combinations of $\mathbf{Q}$ and ω that can be formed from the distributions of $\mathbf{k}_i$ and $\mathbf{k}_f$ given by P_i and P_f. The out-of-plane components of $\mathbf{k}_i$, $\mathbf{k}_f$, and $\mathbf{Q}$ are uncorrelated with the in-plane components and ω. Hence, we can consider transforming from the 2D distributions for the in-plane components of $\mathbf{k}_i$ and $\mathbf{k}_f$ characterized by the constant-probability ellipses in Fig. A4.3 to a 3D distribution characterized by an ellipsoid in $(\Delta\omega, \Delta\mathbf{Q}_\parallel, \Delta\mathbf{Q}_\perp)$ space.

For simplicity, we will consider a symmetric ($\alpha_0 = \alpha_3$, $\alpha_1 = \alpha_2$, $\eta_{\mathrm{M}} = \eta_{\mathrm{A}}$) left-handed W ($\varepsilon_{\mathrm{M}} = -1$, $\varepsilon_{\mathrm{S}} = +1$, $\varepsilon_{\mathrm{A}} = -1$) configuration with $\omega \approx 0$. First, consider a vector $\mathbf{Q}$ defined by $\bar{k}_i$ and $\mathbf{k}_f$, with $\mathbf{k}_f$ ending on the P_f ellipse. A choice of $\mathbf{Q}$ that involves either a pure $\Delta\mathbf{Q}_\parallel$ or $\Delta\mathbf{Q}_\perp$ displacement from $\mathbf{Q}_0$ corresponds to a change in length of $\mathbf{k}_f$, and hence to a change in ω. A positive $\Delta\mathbf{Q}_\parallel$ ($\Delta\mathbf{Q}_\perp$) displacement increases (decreases) k_f, and thus correlates with a negative (positive) value of $\Delta\omega$. Next, consider a $\mathbf{Q}$ defined by $\mathbf{k}_i$ and $\bar{k}_f$. In this case, a positive $\Delta\mathbf{Q}_\parallel$ or $\Delta\mathbf{Q}_\perp$ is correlated with a positive $\Delta\omega$. Thus, one can see by symmetry that there will be no net correlation between $\Delta\mathbf{Q}_\parallel$ and $\Delta\omega$, whereas the triple-axis resolution ellipsoid has one principal axis lying in the $\Delta\mathbf{Q}_\perp$–$\Delta\omega$ plane with a positive slope. Furthermore, for moderate values of θ_{S} the single-axis ellipses tend to have their major axes roughly orthogonal to $\mathbf{Q}_0$, so that the longest axis of the triple-axis resolution ellipsoid is the one in the $Q_\perp$–ω plane. We end up with a picture of an ellipsoid with its longest axis orthogonal to $\mathbf{Q}_0$ and tilted with a negative slope in the $\Delta\mathbf{Q}_\perp$–$\Delta\omega$ plane. A quantitative example is illustrated in Fig. 4.3. The slope in the $\Delta\mathbf{Q}_\perp$–$\Delta\omega$ plane becomes negative for a right-handed spectrometer.

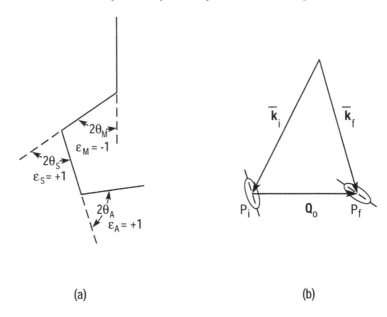

(a)   (b)

Fig. A4.4. Example of a non-W configuration.

In the example above, the spectrometer was assumed to be in the W configuration, as shown in Fig. A4.2. For a left-handed spectrometer (as shown) both the scattering angles $2\theta_M$ and $2\theta_A$ are negative (clockwise rotation). Suppose the analyzer is rotated to scatter in the opposite direction $(2\theta_A \rightarrow -2\theta_A)$ as indicated in Fig. A4.4(a). The ellipsoid corresponding to P_f will also be rotated in the opposite direction, as shown in Fig. A4.4(b). This configuration will result in a different correlation between $\Delta\omega$ and $\Delta\mathbf{Q}$. Under certain circumstances, the orientation of the resolution ellipsoid for this alternative configuration may be preferable to the conventional situation. An example in which it was used to advantage is the study of 1D spin excitations in $KCuF_3$ by Tennant *et al.* (1995).

A4.3 Derivation of the triple-axis resolution matrix

A4.3.1 *Transformation to a common coordinate system*

In order to combine the functions $P_i(\mathbf{k}_i - \bar{\mathbf{k}}_i)$ and $P_f(\mathbf{k}_f - \bar{\mathbf{k}}_f)$ to form the resolution function $R(\omega - \omega_0, \mathbf{Q} - \mathbf{Q}_0)$ following Eq. (A4.3), it is necessary to transform the vectors $\Delta\mathbf{k}_i$ and $\Delta\mathbf{k}_f$ and the matrices A_i and A_f to a common coordinate system defined relative to $\mathbf{Q}_0$ (see Fig. A4.3). We must rotate $\Delta\mathbf{k}_i$ through $90° + \theta_i$ and $\Delta\mathbf{k}_f$ through $90° - \theta_f$, where these rotations are in a negative (clockwise) sense. If we denote quantities rotated into the common coordinate system with a ~ , then

$$\Delta \tilde{\mathbf{k}}_i = \mathsf{O}_i \Delta \mathbf{k}_i, \quad \tilde{\mathsf{A}}_i = \mathsf{O}_i \mathsf{A}_i \mathsf{O}_i^T,$$
$$\Delta \tilde{\mathbf{k}}_f = \mathsf{O}_f \Delta \mathbf{k}_f, \quad \tilde{\mathsf{A}}_f = \mathsf{O}_f \mathsf{A}_f \mathsf{O}_f^T,$$

(A4.40)

where

$$\mathsf{O}_i = \begin{pmatrix} -\sin \theta_i & \varepsilon_{\mathrm{S}} \cos \theta_i \\ -\varepsilon_{\mathrm{S}} \cos \theta_i & -\sin \theta_i \end{pmatrix}, \quad \mathsf{O}_f = \begin{pmatrix} \sin \theta_f & \varepsilon_{\mathrm{S}} \cos \theta_f \\ -\varepsilon_{\mathrm{S}} \cos \theta_f & \sin \theta_f \end{pmatrix}.$$

(A4.41)

One can show that

$$Q_0 \sin \theta_i = \bar{k}_i - \bar{k}_f \cos 2\theta_{\mathrm{S}}, \quad Q_0 \cos \theta_i = \bar{k}_f \sin 2\theta_{\mathrm{S}},$$
$$Q_0 \sin \theta_f = \bar{k}_f - \bar{k}_i \cos 2\theta_{\mathrm{S}}, \quad Q_0 \cos \theta_f = \bar{k}_i \sin 2\theta_{\mathrm{S}}.$$

(A4.42)

These formulas not only define θ_i and θ_f, but can also be used to express $2\theta_{\mathrm{S}}$ in terms of the wave vectors:

$$\cos 2\theta_{\mathrm{S}} = (\bar{k}_i^2 + \bar{k}_f^2 - Q_0^2)/2\bar{k}_i \bar{k}_f.$$

(A4.43)

Thus, one can see that the elements of $\tilde{\mathsf{A}}_i$ and $\tilde{\mathsf{A}}_f$ will be functions of $\bar{k}_i$, $\bar{k}_f$, and Q_0.

A4.3.2 Integration over $\mathbf{k}_i$ and $\mathbf{k}_f$

To obtain the four-dimensional resolution function $R(\omega - \omega_0, \mathbf{Q} - \mathbf{Q}_0)$ from the six-dimensional function $P_i(\mathbf{k}_i - \bar{\mathbf{k}}_i)P_f(\mathbf{k}_f - \bar{\mathbf{k}}_f)$ it is necessary to integrate out two degrees of freedom. Cooper and Nathans (1967) chose to integrate over two components of $\Delta \mathbf{k}_i$; however, that choice tends to obscure much of the symmetry of the problem. Grimm (1984) has demonstrated a more elegant approach that makes the symmetry manifest. The idea is to first transform from the variables $\Delta \mathbf{k}_i$ and $\Delta \mathbf{k}_f$ to $\Delta \mathbf{Q}$ and $\mathbf{x}$. The new variable $\mathbf{x}$ is chosen such that after substituting for $\Delta \mathbf{k}_i$ and $\Delta \mathbf{k}_f$ in the exponential factors in P_i and P_f there are no cross terms between $\Delta \mathbf{Q}$ and $\mathbf{x}$. If we write $\mathsf{M}_i = \bar{k}_i^{-2} \tilde{\mathsf{A}}_i$ and $\mathsf{M}_f = \bar{k}_f^{-2} \tilde{\mathsf{A}}_f$, then the desired transformation is

$$\Delta \mathbf{k}_i = \mathbf{x} - \mathsf{M}_+^{-1} \mathsf{M}_f \Delta \mathbf{Q},$$

(A4.44)

$$\Delta \mathbf{k}_f = \mathbf{x} + \mathsf{M}_+^{-1} \mathsf{M}_i \Delta \mathbf{Q},$$

(A4.45)

where

$$\mathsf{M}_\pm = \mathsf{M}_i \pm \mathsf{M}_f.$$

(A4.46)

To evaluate the integrals in Eq. (A4.3) we start by performing the integration over $\Delta \mathbf{k}_f$, which is easily done by making use of the δ-function for momentum conservation. We then apply the transformation given by Eqs. (A4.44) and (A4.45), with $\Delta \mathbf{k}_f = \Delta \mathbf{k}_i + \Delta \mathbf{Q}$, obtaining

$$R(\Delta \omega, \Delta \mathbf{Q}) = \frac{\hbar^2}{m_n} R_M R_A P_{i0} P_{f0} \exp\left(-\tfrac{1}{2}\Delta \mathbf{Q} G \Delta \mathbf{Q}\right) I(\Delta \omega, \Delta \mathbf{Q}), \qquad (A4.47)$$

where

$$I(\Delta \omega, \Delta \mathbf{Q}) = \int d\mathbf{x} \, \exp\left(-\tfrac{1}{2}\mathbf{x} M_+ \mathbf{x}\right) \delta[\Delta \omega + (\hbar/m_n)(\mathbf{V} \cdot \Delta \mathbf{Q} + \mathbf{Q}_0 \cdot \mathbf{x})], \quad (A4.48)$$

with

$$G = M_i M_+^{-1} M_f = (M_i^{-1} + M_f^{-1})^{-1}, \qquad (A4.49)$$

$$\mathbf{V} = \tfrac{1}{2}\left(\mathbf{K}_0 + M_- M_+^{-1} \mathbf{Q}_0\right), \qquad (A4.50)$$

and

$$\mathbf{K}_0 = \bar{\mathbf{k}}_i + \bar{\mathbf{k}}_f. \qquad (A4.51)$$

To evaluate the integral I it is convenient to work in a coordinate system defined relative to $\mathbf{Q}_0$ (see Fig. A4.3), so that $\mathbf{Q}_0 \cdot \mathbf{x} = Q_0 x_\parallel$. With that choice, one can first integrate over $x_\perp$ and x_z, and then use the δ-function to evaluate the integral over $x_\parallel$. The result is

$$I(\Delta \omega, \Delta \mathbf{Q}) = \frac{m_n}{\hbar Q_0} \frac{2\pi}{\sqrt{(M_+)_{22}(M_+)_{33}}} \exp\left[-\tfrac{1}{2}C\left(\frac{m_n}{\hbar Q_0}\Delta \omega + \mathbf{v} \cdot \Delta \mathbf{Q}\right)^2\right],$$
$$(A4.52)$$

where

$$C = 1/(M_+^{-1})_{11} \qquad (A4.53)$$

and $\mathbf{v} = \mathbf{V}/Q_0$. Putting together all of our results we can finally write $R(\mathbf{Q} - \mathbf{Q}_0, \omega - \omega_0)$ in the form of Eq. (A4.1) with

$$R_0 = \frac{\hbar}{Q_0} \frac{2\pi}{\sqrt{(M_+)_{22}(M_+)_{33}}} P_{i0} P_{f0} R_M R_A, \qquad (A4.54)$$

$$M = \begin{pmatrix} C & v_\parallel C & v_\perp C & 0 \\ v_\parallel C & v_\parallel^2 C + G_{11} & v_\parallel v_\perp C + G_{12} & 0 \\ v_\perp C & v_\parallel v_\perp C + G_{12} & v_\perp^2 C + G_{22} & 0 \\ 0 & 0 & 0 & G_{33} \end{pmatrix}. \qquad (A4.55)$$

Unfortunately, the elements of M are fairly involved functions of $\bar{k}_i$, $\bar{k}_f$, and

Q_0, as well as of $\tan\theta_M$ and $\tan\theta_A$. The only element that has a simple form is G_{33} (corresponding to vertical resolution) which is given by

$$G_{33} = \frac{a_{i,33}a_{f,33}}{\overline{k}_f^2 a_{i,33} + \overline{k}_i^2 a_{f,33}}. \tag{A4.56}$$

Note that in Eq. (A4.1), the matrix M is defined with respect to the 4-vector $\Delta\mathcal{Q}$, with $\Delta\mathcal{Q}_0 = m_n\Delta\omega/\hbar Q_0$. For practical applications, it is generally more convenient to work with $\hbar\Delta\omega$ rather than $\Delta\mathcal{Q}_0$. To achieve this, one can make the transformation

$$\tilde{M}_{00} = M_{00}(m_n/\hbar^2 Q_0)^2, \tag{A4.57}$$

$$\tilde{M}_{0j} = \tilde{M}_{j0} = M_{0j}(m_n/\hbar^2 Q_0), \quad \text{for } j \neq 0. \tag{A4.58}$$

A4.3.3 Characterization of the resolution function

There are a few simple formulas for characterizing the resolution function in terms of the elements of the resolution matrix. The energy width for a constant-Q scan through a dispersionless scattering surface (such as elastic incoherent scattering) is given by

$$\hbar^2\Delta\omega^2 = \frac{\tilde{M}_{22}}{\tilde{M}_{00}\tilde{M}_{22} - \tilde{M}_{02}^2}. \tag{A4.59}$$

The slope of the resolution ellipsoid in the ω–$Q_\perp$ plane is

$$\frac{\hbar\Delta\omega}{\Delta Q_\perp} = \tan\Phi, \tag{A4.60}$$

where Φ is measured relative to the $Q_\perp$ axis, and

$$\tan 2\Phi = \frac{-2\tilde{M}_{02}}{\tilde{M}_{00} - \tilde{M}_{22}}. \tag{A4.61}$$

A4.3.4 Special case: elastic scattering

For elastic scattering we have $\overline{k}_i = \overline{k}_f = k$ and $\theta_i = \theta_f = \theta_S$. For a symmetric spectrometer (i.e., $\alpha_0 = \alpha_3$, $\alpha_1 = \alpha_2$, and $\eta_M = \eta_A$) the matrices M_i and M_f are identical except that their off-diagonal elements have opposite signs. As a result, one finds that G is diagonal, $v_\parallel = 0$, and the resolution matrix simplifies to

$$M = \begin{pmatrix} C & 0 & v_\perp C & 0 \\ 0 & G_{11} & 0 & 0 \\ v_\perp C & 0 & v_\perp^2 C + G_{22} & 0 \\ 0 & 0 & 0 & G_{33} \end{pmatrix}. \tag{A4.62}$$

Thus, under these symmetric conditions, both $\Delta\hat{Q}_\parallel$ and $\Delta\hat{Q}_z$ are principal axes of the resolution matrix. The slope of the major axis of the ellipse in the $\Delta\omega$–$\Delta Q_\perp$ plane can be written

$$\frac{\hbar\Delta\omega}{\Delta Q_\perp} = \frac{\hbar^2 Q_0}{m_{\mathrm{n}}} \tan\Phi', \tag{A4.63}$$

where

$$\tan 2\Phi' = \frac{-2v_\perp}{1 - v_\perp^2 + G_{22}/C}. \tag{A4.64}$$

For the case $\eta_{\mathrm{M}} = 0$, one has $G_{22} = 0$ and $\tan\Phi' = -v_\perp$. Under the further conditions that $\alpha_0 = \alpha_1$ we find that the axis of the ellipse is given by the simple formula

$$\frac{\hbar\Delta\omega}{\Delta Q_\perp} = \frac{\hbar^2 k}{m_{\mathrm{n}}} \cos\theta_{\mathrm{S}}[1 + \tan\theta_{\mathrm{S}}\tan(\theta_{\mathrm{S}} - \theta_{\mathrm{M}})]. \tag{A4.65}$$

(This formula also provides a useful estimate of the slope for finite η_{M}, and $\alpha_0 \neq \alpha_1$.) With $\eta_{\mathrm{M}} = \eta_{\mathrm{A}} = 0$ the constant-probability ellipsoids described by M_i and M_f each collapse in the horizontal plane to a straight line. For $\theta_{\mathrm{S}} = \theta_{\mathrm{M}}$ the two lines are parallel to each other and perpendicular to $\mathbf{Q}_0$. It follows that the width along $\Delta\hat{Q}_\parallel$ of the ellipsoid determined by M (width $\sim 1/\sqrt{G_{11}}$) is zero. For finite η_{M} the longitudinal width becomes finite. For $\theta_i \neq \theta_f$ the resolution ellipsoid rotates so that $\Delta\hat{Q}_\parallel$ is no longer a principal axis.

A4.3.5 Sample mosaic

As discussed in Chap. 4, the signal that reaches the detector is determined by a convolution of the resolution function with the scattering function, $S(\mathbf{Q}, \omega)$, for the sample. One would generally like to measure the intrinsic $S(\mathbf{Q}, \omega)$ for the material under study; however, most real samples have a finite mosaic width, so that what one actually measures is an effective function, $S'(\mathbf{Q}, \omega)$. Let us assume that the sample's mosaic distribution has the form of Eq. (A4.9), where we replace the subscripts M by S. Then the effective scattering function may be written as

$$S'(\mathbf{Q}, \omega) = \int \frac{dQ'_\perp dQ'_z}{Q_0^2} P_{\mathrm{S}}(\mathbf{Q}' - \mathbf{Q}) S(\mathbf{Q}', \omega). \tag{A4.66}$$

Although the resolution function is completely independent of the sample and its orientation, it is often useful to combine the effects of the sample mosaic with the resolution function, and thus work with a modified resolution function R'. This transformation involves setting

$$\int d\omega d\mathbf{Q}\, R(\omega - \omega_0, \mathbf{Q} - \mathbf{Q}_0) S'(\mathbf{Q}, \omega) = \int d\omega d\mathbf{Q}\, R'(\omega - \omega_0, \mathbf{Q} - \mathbf{Q}_0) S(\mathbf{Q}, \omega). \tag{A4.67}$$

Inserting Eq. (A4.66), one finds that

$$R'(\omega - \omega_0, \mathbf{Q} - \mathbf{Q}_0) = \int \frac{dQ'_\perp dQ'_z}{Q_0^2} P_S(\mathbf{Q} - \mathbf{Q}') R(\omega - \omega_0, \mathbf{Q}' - \mathbf{Q}_0). \tag{A4.68}$$

If we write

$$R'(\Delta\omega, \Delta\mathbf{Q}) = R'_0 \exp\left(\tfrac{1}{2}\Delta\mathbf{2} \mathbf{M}' \Delta\mathbf{2}\right), \tag{A4.69}$$

then the modified matrix elements are given by (Werner and Pynn, 1971)

$$M'_{ij} = \begin{cases} M_{ij} - M_{i2}M_{j2}/[M_{22} + (\eta_S Q_0)^{-2}] & \text{for } i, j = 0, 1, 2 \\ M_{33} - M_{33}^2/[M_{33} + (\eta'_S Q_0)^{-2}] & \text{for } i = j = 3 \end{cases} \tag{A4.70}$$

and

$$R'_0 = R_0 \left[\left(1 + M_{22}\eta_S^2 Q_0^2\right)\left(1 + M_{33}\eta_S'^2 Q_0^2\right)\right]^{-1/2}. \tag{A4.71}$$

A4.4 The ellipsoid of constant probability

To simplify the notation in this section, we drop the symbol Δ and let $\hbar \to 1$, so that, for example, $\hbar\Delta\omega \to \omega$. Using this notation, the ellipsoid of constant probability for the three coupled coordinates in the resolution function can be written

$$\tilde{M}_{00}\omega^2 + \tilde{M}_{11}Q_\parallel^2 + \tilde{M}_{22}Q_\perp^2 + 2\tilde{M}_{01}\omega Q_\parallel + 2\tilde{M}_{02}\omega Q_\perp + 2\tilde{M}_{12}Q_\parallel Q_\perp = c, \tag{A4.72}$$

where c is a constant. To describe the 50% probability contour, one sets $c = 2\ln 2$. The cross section of the ellipsoid with the plane described by any pair of coordinates is obtained by setting the third coordinate to zero in the equation above. For example, the cross section in the ω-$Q_\parallel$ plane is given by

$$\tilde{M}_{00}\omega^2 + \tilde{M}_{11}Q_\parallel^2 + 2\tilde{M}_{01}\omega Q_\parallel = c. \tag{A4.73}$$

The formula for the projection of the ellipsoid onto a particular plane can be derived by integrating the resolution function over the third coordinate. For the ω-$Q_\parallel$ plane the result is

$$\left(\tilde{M}_{00} - \frac{\tilde{M}_{02}^2}{\tilde{M}_{22}}\right)\omega^2 + \left(\tilde{M}_{11} - \frac{\tilde{M}_{12}^2}{\tilde{M}_{22}}\right)Q_\parallel^2 + 2\left(\tilde{M}_{01} - \frac{\tilde{M}_{02}\tilde{M}_{12}}{\tilde{M}_{22}}\right)\omega Q_\parallel = c. \tag{A4.74}$$

Similarly, for ω–$Q_\perp$ one has

$$\left(\tilde{M}_{00} - \frac{\tilde{M}_{01}^2}{\tilde{M}_{11}}\right)\omega^2 + \left(\tilde{M}_{22} - \frac{\tilde{M}_{12}^2}{\tilde{M}_{11}}\right)Q_\parallel^2 + 2\left(\tilde{M}_{02} - \frac{\tilde{M}_{01}\tilde{M}_{12}}{\tilde{M}_{11}}\right)\omega Q_\parallel = c,$$

(A4.75)

and for $Q_\parallel$–$Q_\perp$,

$$\left(\tilde{M}_{11} - \frac{\tilde{M}_{01}^2}{\tilde{M}_{00}}\right)\omega^2 + \left(\tilde{M}_{22} - \frac{\tilde{M}_{02}^2}{\tilde{M}_{00}}\right)Q_\parallel^2 + 2\left(\tilde{M}_{12} - \frac{\tilde{M}_{01}\tilde{M}_{02}}{\tilde{M}_{00}}\right)\omega Q_\parallel = c.$$

(A4.76)

To plot an ellipse, it is convenient to work in polar coordinates. For example, to plot Eq. (A4.73) one can set $\omega = r\cos\theta$ and $Q_\parallel = r\sin\theta$. Solving for r yields

$$r = [c/(\tilde{M}_{00}\cos^2\theta + \tilde{M}_{11}\sin^2\theta + 2\tilde{M}_{01}\sin\theta\cos\theta)]^{1/2}.$$

(A4.77)

References

Cooper, M. J. and Nathans, R. (1967). *Acta Cryst.* **23**, 357.

Dorner B. (1971). *J. Appl. Cryst.* **4**, 185.

Dorner, B. (1972). *Acta Cryst. A* **28**, 319.

Grimm, H. (1984). *Nucl. Instrum. Methods* **219**, 553.

Riste, T. and Otnes, K. (1969). *Nucl. Instrum. Methods* **75**, 197.

Sailor, V. L., Foote, Jr, H. L., Landon, H. H., and Wood, R. E. (1956). *Rev. Sci. Instrum.* **27**, 26.

Tennant, D. A., Nagler, S. E., Welz, D., Shirane, G., and Yamada, K. (1995). *Phys. Rev. B* **52**, 13381.

Werner, S. A. and Pynn, R. (1971). *J. Appl. Phys.* **42**, 4736.

Index